Topics in Current Physics

16

Topics in Current Physics Founded by Helmut K. V. Lotsch

1 **Beam-Foil Spectroscopy**
Editor: S. Bashkin

2 **Modern Three-Hadron Physics**
Editor: A. W. Thomas

3 **Dynamics of Solids and Liquids by Neutron Scattering**
Editors: S. W. Lovesey and T. Springer

4 **Electron Spectroscopy for Surface Analysis**
Editor: H. Ibach

5 **Structure and Collisions of Ions and Atoms**
Editor: I. A. Sellin

6 **Neutron Diffraction**
Editor: H. Dachs

7 **Monte Carlo Methods** in Statistical Physics
Editor: K. Binder

8 **Ocean Acoustics**
Editor: J. A. DeSanto

9 **Inverse Source Problems** in Optics
Editor: H. P. Baltes

10 **Synchrotron Radiation** Techniques and Applications
Editor: C. Kunz

11 **Raman Spectroscopy** of Gases and Liquids
Editor: A. Weber

12 **Positrons in Solids**
Editor: P. Hautojärvi

13 **Computer Processing of Electron Microscope Images**.
Editor: P. W. Hawkes

14 **Excitons** Editor: K. Cho

15 Physics of **Superionic Conductors**
Editor: M. B. Salamon

16 **Aerosol Microphysics I** Particle Interactions
Editor: W. H. Marlow

17 **Solitons**
Editors: R. Bullough, P. Caudrey

18 **Magnetic Electron Lens Properties**
Editor: P. W. Hawkes

19 Theory of **Chemisorption**
Editor: J. R. Smith

20 **Inverse Scattering Problems** in Optics
Editor: H. P. Baltes

21 **Advances in Coherent Nonlinear Optics**
Editors: M. S. Feld and V. S. Letokhov

22 **Electromagnetic Theory of Gratings**
Editor: R. Petit

Aerosol Microphysics I
Particle Interaction

Edited by W. H. Marlow

With Contributions by
J. R. Brock H. W. Chew J. D. Doll M. Kerker
W. H. Marlow P. J. McNulty

With 35 Figures

Springer-Verlag Berlin Heidelberg New York 1980

William H. Marlow, Ph. D.

Brookhaven National Laboratory, 51 Bell Avenue, Upton, NY 11973, USA

ISBN-13: 978-3-642-81426-6 e-ISBN-13: 978-3-642-81424-2
DOI: 10.1007/978-3-642-81424-2

Library of Congress Cataloging in Publication Data. Main entry under title: Aerosol microphysics, I. (Topics in current physics; v. 16). Bibliography: p. Includes index. 1. Aerosols. 2. Particles. I. Marlow, William H. II. Brock, J. R. III. Series. QC882.A35 541.3'4515 79-28486

Softcover reprint of the hardcover 1st edition 1980

Offset printing and bookbinding: Konrad Triltsch, Graphischer Betrieb, Würzburg.
2153/3130-543210

Preface

The suggestion by Dr. Franklin S. Harris, Jr., that these books be written arose pursuant to the editor's plaints that despite the implicitly or explicitly acknowledged importance of both aerosols and particulate matter in innumerable domains of technology and human welfare, investigations of these subjects were generally not supported independently of the narrowest conceivable domains of their applications. Frank Harris, who has long been a contributor in one of the important domains of aerosol macrophysics, atmospheric optics, challenged the editor to elaborate his views. Ideally, they would have taken the form of a monograph; however, there is as yet an insufficient body of information to present a unified treatment. At the same time, substantial efforts are in progress in the component fields to hold the promise for the emergence of unifying elements which will eventually facilitate their presentation to be made with a high degree of integrity.

There are numerous pertinent and systematic tie-ins between project-oriented aerosol work and basic physical investigations which are themselves quite closely akin to much classical and current work in physical science. The most significant aspect of these tie-ins is their potential for making substantial contributions to the functional needs of the applications areas while stimulating significant questions of basic physics. For this to be possible, it is necessary that the most relevant areas of physics be identified in such a manner as to make clear their relevance for aerosol-related studies and vice versa. Thus, these books differ from perhaps all others in that they are actually physics books whose topics are chosen for the central roles they play in that physical system commonly known as the aerosol. Fundamental, specifically aerosol questions have not been discussed in detail because of the editor's conviction that their complete description lies in the application of one or more of the domains identified in these books. The extent to which these books not only serve to identify the classes of physical questions that must be asked before any aerosol system can be measured, modelled, or otherwise described but also help to provide the link between selected basic physical disciplines and their applications will ultimately determine their contributions to both aerosol science and to physics.

Upton, NY, November 1979 *W.H. Marlow*

Contents

List of Contributors

Brock, James Rush

Department of Chemical Engineering, University of Texas, Austin, TX 78712, USA

Chew, Herman W.

Department of Physics, Clarkson College of Technology, Potsdam, NY 13676, USA

Doll, Jimmie Dave

Los Alamos Scientific Laboratory, Los Alamos, NM 87545, USA

Kerker, Milton

Department of Physics, Clarkson College of Technology, Potsdam, NY 13676, USA

Marlow, William H.

Department of Energy and Environment, Brookhaven National Laboratory, Upton, NY 11973, USA

McNulty, Peter J.

Department of Physics, Clarkson College of Technology, Potsdam, NY 13676, USA

1. Introduction: The Domains of Aerosol Physics

W. H. Marlow

Between matter-in-bulk and isolated molecules or atoms, there is virtually a continuum of material aggregations. If these aggregates exist in the gas phase, which they do when they are not immobilized upon or within bulk matter, then they constitute the particulate (either liquid or solid) component of the condition of matter generally referred to as an aerosol. Among researchers in basic and applied studies in both the sciences and the fields of engineering and technology, there is at best only mixed understanding of the generality of the aerosol condition of matter in natural and man-made systems. Generally speaking, there is no appreciation that as a consequence of this generality, common physical phenomena underlie many of their questions. Thus, the studies of aerosols or gas-borne particulate matter have remained fragmented gaining their identities from the numerous basic and applied disciplines in which they arise. The result of this fragmentation for both these aerosol-related studies and relevant fields is their development without regard for the accumulated wisdom and techniques of the other fields and the cost is both duplication of efforts and missed opportunities for progress.

This introductory chapter addresses the concept of the aerosol condition of matter first by explaining how it arose and then by delineating the physical questions which are common to all such systems. Section 1.1 reviews possible definitions of "aerosol" and Sect.1.2 discusses the division among the component studies of aerosol physics which is followed in these volumes. The discussion of aerosol microphysics in Sect.1.2.2 establishes its relationship to the conventional investigations of basic physics. To be of value both to the basic scientist seeking the context of his work in other natural sciences and in technology and conversely to the engineer or natural scientist curious of how his domain of interest is related to aerosol microphysics, Sect.1.3 briefly discusses examples of where aerosols arise.

1.1 The Aerosol Concept

Various definitions have been given to describe the systems of particulate matter dispersed in the gas phase. Originally, the term "aerosol" was coined in analogy to hydrosol to describe fine smokes [1.1]. However, the analogy was soon realized not to be a good one since an aerosol is generally a nonequilibrium state. Refinements and subclassifications based on popular terminology such as fumes, dusts, mists, smokes, etc. [1.1,2] and their means of generation [1.2] have also been given. None of these appear satisfactory, however, because aerosols encountered in the broadest varieties of studies may be mixtures arising from numerous sources thus vitiating the usefulness of the source classification. The alternative approach is the definition in terms of particle physical properties [1.3] which leaves open the questions of generation, interactions, and evolution. These properties require the particle Reynolds number, Re_i ($Re_i = d_g q_i R_i / \mu_g$ where d_g denotes the gas density, q_i the average particle drift velocity, R_i is particle radius, and μ_g is gas viscosity) to be less than unity and the surface to volume ratio to be much greater than 1000 cm^{-1}. Even this definition leaves something to be desired. Frequently particle radius is not defined due to nonsphericity of the particle, and its operational definition in terms of some kind of equivalent radius (e.g. aerodynamic-equivalent diameter, optically equivalent sphere, electric-mobility equivalent radius) is entirely dependent upon measurement method. Some of the most interesting and important properties of aerosol particles from the viewpoints of both their basic science and their effects or application are attributable to aspects of their nonsphericity. Similarly, the surface-to-volume definition is overly restrictive because it effectively eliminates particles over 10 μm. Such particles can remain in suspension for extended periods of time under a variety of conditions and may constitute a very important type of aerosol, the fluidized bed [1.2]. There is also the question of how to define the surface, for example, for porous particles or for molecular clusters. Due to these difficulties in the formulation of an unambiguous definition, the generic term, aerosol, will be taken here to include all gas-phase particulate suspension [1.4] without further elaboration either of particle or gas properties (i.e., large molecules should be included as particles) or their boundary conditions such as external or internal fields.

1.2 Aerosol Physics

In the systematic study of aerosol physics it is useful to recognize at division into *aerosol microphysics* and *aerosol macrophysics*. Phenomena involving the formation, transport, and interaction of one or two (or, more generally, a small number) of particles or of these particles with the gas are grouped under microphysics. In aerosol macrophysics, bulk, collective, or cooperative properties of the particle-gas system taken as a whole are treated. An appropriate analogy can be drawn by recognizing that the relationship of aerosol microphysics to aerosol macrophysics is similar to the relationship between molecular and solid-state physics.

1.2.1 Macrophysics

Aerosol macrophysics can be thought of as encompassing at least the studies of *communitive aerosol generation, optical turbidity, filtration,* and *aerosol field dynamics* (e.g., aerosol dynamics in turbulence, acoustic, electrical, etc. fields).

Communitive particle generation is that class of processes by which "bulk" material is mechanically broken into particles. It may be done by a variety of means, but in many cases, the production of a single particle has scant significance since particles frequently fuse, thereby limiting the degree of subdivision of the original material. Clearly, questions of the structure of materials, interfacial science, and the theory of random processes are significant [1.1,3,5].

Optical turbidity and radiative transport in the present context are concerned with the relationship between light incident upon an aerosol and that emergent from it. Since even elastic scattering of light by a single particle is dependent upon wavelength, particle composition, shape, structure, and size, the scattering by a *polydisperse* (particles of many sizes simultaneously present) *aerosol* of a broad spectrum of wavelengths becomes a study of its own, with the optical interactions of a single particle useful only as a starting point [1.6-8,51].

Filtration, in the most general sense, may be defined as the removal of particles from the aerosol. This occurs either by their attachment to nonaerosol media (walls, vegetation, "fabric filters", etc.) or to larger particles which are subsequently removed. Since particle transport in the gas is intimately involved, a characterization of the gas flow *field* and the detailed mechanisms of particle kinetic theory near a surface must be invoked. Classically, filtration was treated as the simple adhesion of a single particle to a surface. However, it is now known that after the first particles adhere, subsequent ones tend to be captured by the initial ones to form chains. Impaction of a large particle upon such a chain or other break-off processes can cause resuspension. Thus, filtration is dependent upon properties of the aerosol and gas as a whole [1.9,10].

Aerosol field dynamical studies comprise those investigations of the factors that govern the (spacially) large-scale behavior of the particles. Turbulent transport of particles is the motion of particles in turbulent flow fields. Since the response of a particle to this flow is, by definition, dependent upon its aerodynamic properties, differential motion between particles of differing sizes can greatly influence coagulation and, therefore, the aerosol size distribution.

Clearly, the turbulence field is also of central importance in filtration since it affects the transport of the particle to the vicinity of the filter surface [1.3,4]. A second variety of field is the acoustic field which affects the short-range motion of all the particles. In this case, it is largely the size-dependent local motion of particles (due to the acoustically caused density gradients) which is perturbed and which in turn is used to enhance particle coagulation [1.11]. A third example of particle field dynamics is the behavior of particles in an electromagnetic field. Dielectrophoresis and simple electrostatic attraction cause particles to move. Since a wide variety of particle mobilities is often involved, with simultaneous transport, charging, and coagulation possible; the flow, electromagnetic, and particle size distribution fields are simultaneously involved [1.12, 13].

In none of these macrophysical studies does the isolated behavior of a small number of particles and gas molecules have any meaning. They are collective phenomena whose study is defined by the "extensive" properties of their environment (e.g., flow fields), particle size distribution, gas composition etc.

1.2.2 Microphysics

Any division of aerosol macrophysics must ultimately be either formulated in terms of aerosol microphysics or at least consistent with it. In many cases, details of the "microscopic" physical properties of particles in a gaseous medium contribute little insight into the overall process. For example, charging of spheres which are somewhat larger than a gas molecular mean free path is completely defined in terms of macroscopic parameters. Conversely, smaller particle charging is poorly defined without a detailed accounting of the charging-ion's transport in the vicinity of the particle (see [1.14] for a discussion of these points in the case of diffusion charging).

Since the impact of, and concerns with aerosols have usually been with their large scale properties in general and most frequently with only the larger particles, the role of aerosol microphysics has generally been poorly defined or ignored. In recent years, not only have basic physics and chemistry evolved beyond their classical divisions of bulk and completely dispersed states, but also studies in the fields discussed in Sect.1.3 have increasingly become recognized as incomplete by omission of microphysical considerations. Therefore, identification of the subjects in which the research in the basic sciences converges with the needs of the applied sciences is desirable for their systematic development.

The division of physical studies herein termed "aerosol microphysics" is of course somewhat arbitrary. The objective of such a division is to delineate common topics of research investigation and application. In some cases such as photophoresis [1.15], several of these divisions apply in a clearly microphysical problem. Nevertheless, they do identify what are probably the requisite domains of physics for the description of all aerosol phenomena, and they are introduced in the following:

Kinetic theory is the most thoroughly developed subject in so far as its application to aerosols are concerned. Several treatises largely devoted to it [1.2,3] are available. In a sense, the subject treats: 1) the action of the gas on the motion of the particle (i.e., particle transport), and 2) the effect of the particle upon the gas as in all *surface accommodation processes*. Clearly, 1) and 2) are intimately related. A critical quantity in all such considerations is the size of the particle relative to the gas molecular mean free path. When the particle is of comparable magnitude to or smaller than the mean free path, its transport properties must account for the discreteness of the gas. For molecules interacting with a particle, the questions of their motions in the gas prior to impact and upon leaving the surface and entering the gas phase determine both molecular transport processes to and from the particle and the transport of other physical quantities or species in the gas. Chapter 2 reviews recent work in the kinetic theory of aerosols in general, while Chap.3 presents a model for solid surfaces from which general accommodation properties can be calculated.

Optical interactions of particles include elastic and inelastic scattering and spontaneous and stimulated emission of light. For most of these processes, the ratio of wavelength to particle dimension is the critical parameter which determines the appropriate law of physical interaction for either classical or quantum-mechanical treatments. Considerable work has been done on the classical elastic scattering of light from translucent spheres because of its great importance in numerous applied studies. Conventionally, when such a particle was absorptive, the fate of the energy not elastically scattered was not considered, despite the fact that Raman and fluorescence processes could be responsible for the inelastic component of the scattering and, therefore, lead to radiation which was wavelength- or phase-shifted from the incident beam. Chapter 4 presents a classical description of this reradiation. A related problem is the thermal (spontaneous) emission from small particles [1.16,17]. In conventional theory, the Planck distribution of cavity radiation modes is assumed. Two fundamental problems arise in the application of this law to small particles: 1) the "small hole" construct (by which the radiation escapes from the cavity with perfectly internally reflecting walls) ceases to be meaningful for a particle whose dimensions are comparable to the wavelength of radiation in question; 2) the Planck distribution itself is only valid for "infinite" bodies, an approximation that fails completely for particles comparable to a thermal wavelength. For sufficiently small particles, full accounting must also be made for the effects of surface scattering of conduction electrons and the consequent alteration of the particle's electric susceptibility [1.18] as well as alteration of the plasmon structure.

Aerosol interaction forces are due to considerations of both kinetic theory (i.e., gas dynamics) and dynamics (electromagnetic interactions and others which are microscopically independent of the gas molecules). The former lies clearly in the domain of kinetic theory and has been discussed at length in the literature, while the latter is generally treated only when excess charge is present, though van der Waals forces can be important. The critical consideration in assessing interaction forces is the balance between the influence of kinetic theory and dynamics. In perhaps the most important cases, the domains of influence of the dynamics upon particle (or molecular) trajectories lie well within a gas molecular mean free path of the particle surface at room temperature and pressure. However, quantitative treatment requires full accounting of the particle "Brownian" motion in the density gradient near a surface and in the field gradient in that region. Since the interaction potential is ultimately responsible for particle behavior, characterization of the potential energy is important even for qualitative considerations. When charge or charge separation is present on a particle, it dominates the interaction. More commonly, though, the particles are neutral with probably no permanent multipole moments. In these cases, the van der Waals forces govern interactions close to the particles. Since these forces are dependent upon particle chemical composition, internal structure, and morphology, their effect must be assessed for each individual species of interacting particle. Chapter 5 reviews this general field with emphasis on developements arising from the Lifshitz theory for van der Waals forces among solids or liquids.

Homogeneous nucleation is the formation of the condensed phase (particles) from purely gaseous molecules. If only a single molecular species is involved, the process is termed homomolecular, while it is called heteromolecular when more than one such species participates. Aspects of homogeneous nucleation depend to a great extent upon collision rates; this leads to highly mixed results upon treatment by kinetic theoretic means. Undoubtedly, any ultimate description will necessitate details not only of kinetics but also of dynamics and microparticle microphysics to account for the rates and structure of critical (i.e., stable) cluster formation.

Aerosol thermodynamics must account for the Kelvin effect, the rising of the equilibrium vapor pressure of a substance over a curved surface of its condensate relative to that vapor pressure over the flat surface. In this case two problems arise: the lack of definition of surface tension as particle size diminishes and the extension of the theory of phase equilibria in general and the Kelvin effect in particular to include multiple molecular components. Numerous effects of these thermodynamic considerations arise, as in particle transport due to chemical composition gradients in the gas phase.

In the aerosol context, *heterogeneous processes* refer to both the accretion and transformation of molecular species on a preexisting particle or gas-phase ion (i.e., a species heterogeneous to the strictly gas phase). As such it is intimately related to both thermodynamics and surface accommodation. The study of the physics and chemistry of surfaces is a field which has come into its own as a distinct discipline in recent years. It is well known that the structure of a material is essentially that of the bulk matter at depths below two or three molecular layers from the surface. The question becomes blurred if molecular clusters and solids of only five or ten molecular diameters in a characteristic dimension are considered. In effect, then, small particles may be viewed as "three dimensional" surfaces implying that the study of their surface interactions will have unique aspects not to be expected of either the surfaces of bulk media or of molecules.

The *physics of microparticles* [1.19] is the final component of this somewhat arbitrary division of aerosol microphysics into distinct fields of study. The inclusion of three dimensional finitude in the study of condensed systems raises numerous questions including some mentioned explicitly above in the discussions of optical interactions and heterogeneous process. Because the small-particle surface-to-volume ratio may be significant, modifications to physical pictures based on the bulk state are required. For example, the enhanced importance of surface phonons decreases the exponent in the temperature dependence of the Debye specific heat of microparticles. Other modifications concern the electronic structure of particles and, therefore, their dielectric susceptibilities. For clusters, quantization of the conduction-electron-level structures should occur due to their confinement. For larger small particles, modification of light scattering has been observed which can be understood in terms of an alteration of the dielectric susceptibility caused by the reduction of conduction-electronic mean free paths due to scattering from the surface of the particle. In addition to their central roles in photon processes and thermal transport, the modification of these physical parameters of matter from what they are for the bulk material is of crucial importance in particle van der Waals forces.

1.3 The Scope and Impact of Aerosols: Selected Fields

Since aerosols or *highly dispersed matter* are a natural material condition, their occurrence is extremely common. Questions in innumerable fields ranging from astrophysics to combustion technology require the studied treatment of aerosols as they enter into various aspects of those concerns. Nevertheless, in the spirit of any truly basic science, the spectrum of possible questions concerning aerosols and arising from applications can be reduced to the relatively limited number of underlying possibilities discussed in Sect.1.2. Thus, research results in one of these fields may be expected to hold implications for diverse applications. To illustrate how this may occur for microphysics, examples selected from several diverse fields are discussed in this section.

1.3.1 Combustion Technology

In diesel engines, oil burners, coal combustors, etc., fuel is delivered to the combustion region as particles or aerosols. The control of their dispersal and size distribution upon entrance into the combustion region is a matter of aerosol thermodynamics [1.20].

For heterogeneous fuels such as coal-oil mixtures, control of the oil's potential for coating the coal particles requires treatment of its interaction energy [1.21].

In the combustion of small particles, several microphysical questions arise. Thermodynamics of volatile fuel components is important in vaporization; kinetic theory determines rates of all gas-phase transport, and accommodation coefficients; and heterogeneous processes are important for the surface chemistry. Particle optics and microphysics govern both thermal distribution within, and radiative dissipation from particles as well as playing a role in photochemical processes.

Soot, flyash, and other particulate matter are formed in the combustion process [1.22]. Homogeneous nucleation of carbon and other incompletely burned fuel constituents is considered the most likely formation mechanism of soot. Flyash is formed of the noncombustible fuel elements and may evolve from many microphysical causes including heterogeneous processes and coagulation (i.e., interaction forces and kinetic theory) [1.23].

It is well known that a significant part of the energy produced in combustion is carried from the flame or combustion zone by the soot and flyash particles. This energy is then lost by the radiation and conduction. Optical interactions and particle microphysics are central to the radiative transfer, and kinetic theory enters for the thermal conduction processes [1.6,24].

1.3.2 Industrial Processes and Techniques

Wherever particles are involved, the problem of fouling, or deposition on surfaces arises. In the case of heat exchangers for combustion gases, thermal transfer efficiency may be drastically reduced by the deposition of the relatively highly insulating soot and flyash particles. In other contexts sulfuric acid and other corrosive vapor droplets diffuse or impact upon conduit surfaces thereby shortening their useful lifetime [1.22,25]. In all of these cases, numerous questions of kinetic theory arise including all the phoretic forces. In addition, particle interaction forces are ultimately responsible for delivery to the surface in question.

To ameliorate fouling problems as well as other problems involving particles, such as erosion and atmospheric emissions, the technology of hot and cold gas clean-up has arisen [1.26]. Techniques involving centrifugal separations (cyclone), electrical charging and field migration (electrostatic precipitation), preconditioning by growth, agglomeration, scavenging (acoustic and water-spray treatment), and other means to facilitate particle removal, have been proposed or are in use [1.27]. A variety of aerosol microphysical domains enter into these techniques including kinetic theory, interaction forces, heterogeneous processes, thermodynamics, and microphysics.

In recent years there has been a growing realization that clusters and fine particles below 10 nm are far more catalytically active than larger particles or surfaces [1.28]. Based upon this information, matrix-isolated metallic clusters [1.29] and bimetallic species [1.30] have been generated for catalytic work. Heterogeneous processes and kinetic theory are involved in their catalytic function and, in addition, interaction forces play a role in their generation.

1.3.3 Medicine and Health

Inhaled particulate matter is susceptible to capture by the body in the respiratory system. Depending upon the particle's size, this may occur anywhere from the nose or mouth for the largest particles down to the lung's alveoli for the smallest (under 200 nm). Since the respiratory system presents an environment to the aerosol that is quite different from that outside the body, particle growth and transformation frequently occur, complicating the analysis of deposition mechanisms [1.31,32]. Aerosol microphysical domains of relevance include the thermodynamics of particle growth, kinetic theory, interaction forces, and some aspects of homogeneous nucleation theory.

1.3.4 Occupational Hygiene and Safety

One of the constant concerns of numerous industries where either aerosols or toxic vapors are produced is the effects of these upon workers. A variety of health problems arise because of the workers' inhaling gases and aerosols of irritant or toxic fibers which are produced in the normal course of activities. A less obvious hazard arises from the attachment of trace molecular species to particles. If these trace species are radioactive or chemically toxic, the particles, which effectively concentrate them, provide a vehicle for delivery deep within the body causing a health hazard far beyond that suggested by their gross molecular concentration in the air [1.31,32]. Kinetic theory, thermodynamics, and interaction forces, all enter into descriptions of the attachment process.

Dust explosions [1.33] present an industrial hazard in the grain milling and other industries where combustible airborne particles are produced. Describing in detail the conditions under which these explosions can occur requires the aerosol microphyiscs of heterogeneous processes, kinetic theory, and microparticle microphysics.

1.3.5 Planetary Atmospheres

Since particles are ubiquitous in planetary atmospheres [1.34], they play roles in most atmospheric processes.

Atmospheric electricity [1.35,36] is concerned not only with lightning, but generally with the status of the "earth-sky" electric field gradient. The free charges responsible for modifying it are generated primarily by collisional ionization processes due both to decay products of naturally occurring radioactive elements at the surface of the earth and to cosmic rays. These ions rapidly cluster water and other trace gas molecules that become cluster ions which have high electrical mobilities and are the species principally responsible for atmospheric conductivity. However, in the presence of aerosols, this conductivity is reduced due to the capture of ions by particles which have much lower electric mobilities. Thus, particles play a crucial role in maintaining the atmospheric electric field gradient, and their interaction forces, thermodynamics, microphysics, and kinetic theory are involved in the description of these atmospheric electrical phenomena.

Cloud formation and dynamics is a subject which requires considerable information from almost all the areas of aerosol microphysics [1.37,18]. Thermodynamics determines aerosol growth to cloud droplet size, and electrical processes in clouds may play a role in the onset of precipitation. Clouds may scavenge the atmosphere of aerosols through capture of particles by cloud droplets, with the rates and mechanisms described by subtleties of their interaction forces and aspects of kinetic theory.

A problem of overriding importance for planetary atmospheres as a whole is that of the albedo, the ratio of the energy radiated to space by the planet to that incident upon it [1.39]. Particles and cloud droplets are responsible for reflection back to space of a part of the incident energy, while they play roles in the capture and thermalization of the thermal emissions of the "solid" earth. In this latter regard, particles may be particularly important in the question of stratospheric heating [1.40,41]. Besides particle interaction with light, microphysics and kinetic theory may also be important in understanding aspects of the thermal transfer problem.

1.3.6 Air Pollution

The separation of "air pollution" from "planetary atmosphere" is an arbitrary one which arises in recognition of those processes known specifically to be affected or important to human activities in addition to natural ones [1.42].

One example that has received considerable attention in recent years is the origin and fate of atmospheric sulfate particles. There is reason to believe that SO_2 gas released in one region appears 100 km and farther from its source in the form of sulfuric acid and other sulfate-containing particles. Both homogeneous and heterogeneous nucleation are of clear importance in its initial formation.

Atmospheric visibility can be greatly decreased by primary particulate emissions from industrial sources and by particles generated or grown in the atmosphere in addition to naturally occurring causes. Since the visibility reduction is a function of particle size distribution, aerosol thermodynamics and interaction forces are involved in addition to light scattering.

A final example of air pollution concerns is that of the fate of particles. Not only may they be removed by precipitation under clouds, but they attach to surfaces and vegetation. Interaction forces and a variety of aspects of kinetic theory are involved in such processes.

1.3.7 Astrophysics: Interstellar Matter

As they cool following emission from stars, the elementary species (ions and atoms) form molecules and eventually particles via homogeneous nucleation. These particles and perhaps others originating in other ways provide surfaces for heterogeneous processes facilitating or causing the formation of molecular species not possible by purely isolated molecular reactions. Particle optical interactions are partly responsible for both their dispersal in space and for stellar obscuration [1.43].

1.3.8 Relationship to Aspects of Materials Sciences, Biophysics, and Solar Collector Technology

As is clear from the material discussed in Sect.1.2, the divisions of aerosol microphysics are not of unique relevance only to aerosols but are rather a refinement of more general fields of research. It is, therefore, of interest to note that exactly these same categories of research are necessary in other, nonaerosol related work. A few of the innumerable areas are mentioned.

One of the problems with solar-radiation heat collectors has been their inefficient collection of the short-wavelength component of the solar spectrum. When a conventional metallic collector is coated with very fine particles which are absorptive at short wavelengths, the resultant heat generated in the particles can be dissipated by conduction to the collector [1.44]. Both particle optical interactions and microphysics play roles in the design of such particles, and interaction forces are useful for their immobilization.

In biophysics, the questions of cellular and membrane interactions are important. Here, particle interaction forces are very similar, and much the same type of treatment is of relevance in both cases [1.45-47].

In materials science, both the preparation of composite materials and the conditions of use, e.g., catalysts, require access to methods of value for aerosol microphysics. In the former case, aerosol interaction forces play a role not only in understanding the agglomeration properties of the components of the composite, but also in describing the potential for spreading of the interstitial material on the included solid phase. In the latter case, questions of heat dissipation or absorption involve both kinetic theory and optical interactions.

1.3.9 Aerosol Measurement Methodology

Ultimately, the measurement of anything must be either insensitive to variable properties other than those being quantified or it must involve precise descriptions of the relationship of those detected secondary properties with the primary property of interest. For aerosols, this presents a problem of extreme difficulty in the general case because of thermodynamic, aerodynamic, optical, surface, and other variable properties which vitiate the utility of a single method [1.48]. For example, mass, number, and their distribution among particles are primary properties of overriding importance, and numerous methods are used for their measurement [1.49,50]. However, essentially all of them either sense or are dependent upon a secondary property whose connection to the primary property is unclear and sometimes misleading. For instance, the aerodynamic separation methods such as impactors ultimately provide a sample whose mass can be determined without interference. Shape and mass distribution factors, which are not necessarily of any intrinsic importance, are known to obscure the meaning of particle numbers derived from the mass measurements by assuming "aerodynamic equivalent diameters" for the collected particles. Extension of impaction methods to very small particles requires reduced pressures on the impaction side where volatility of constituents of the particles can change their size and composition thereby rendering the meaning of the basic measurement questionable. Improvement of the characterization of the relationship between measurement procedure and measured property is an undertaking to which all phases of aerosol microphysics contribute.

References

1.1 H.L. Green, W.R. Lane: *Particulate Clouds: Dusts, Smokes and Mists*, 2nd ed. (Van Nostrand, Princeton 1964)

1.2 N.A. Fuchs: *The Mechanics of Aerosols*, transl. from Russian by R. E. Daisley, M. Fuchs (Macmillian, New York 1964)

1.3 G.M. Hidy, J.R. Brock: *The Dynamic of Aerocolloidal Systems* (Pergamon Press, Oxford 1970)

1.4 S.K. Friedlander: *Smoke, Dust, and Haze: Fundamentals of Aerosol Behavior* (Wiley, New York 1977)

1.5 A. Kolmogorov: Dokl. Akad. Nauk SSSR, *31*, 99 (1941)

1.6 H.C. Hottel, A.F. Sarofim: *Radiative Transfer* (McGraw-Hill, New York 1967)

1.7 R.M. Goody: *Atmospheric Radiation* (Oxford University Press, London 1964)

1.8 H.C. van de Hulst: *Multiple Light Scattering* (Academic Press, New York 1979)

1.9 C. Orr, Jr., E.Y.H. Keng: "Sampling and Particle Size Measurement" in *Handbook on Aerosols*, ed. by R. Dennis (National Technical Information Service, Springfield, Va. TID-26608, 1976) Chap.5

1.10 M. Lippman: "Filter Media and Filter Holders for Air Sampling", in *Air Sampling Industruments*, 5th ed. (American Conference of Governmental Industrial Hygienists, Cincinnati 1978)
1.11 D.J. Shaw: "Acoustic Agglomeration of Aerosols": in *Recent Developments in Aerosol Science*, ed. by D.J. Shaw (Wiley, New York 1978) Chap.13
1.12 H.A. Pohl: *Dielectrophoresis* (Cambridge University Press, Cambridge 1978)
1.13 J.R. Brock: J. Appl. Phys. *41*, 1940-1944 (1970)
1.14 W.H. Marlow: J. Colloid Interface Sci. *64*, 543-548 (1978)
1.15 M.S. Arnold, L. Rozenshtein: Pub. Chem. Phys. Letters *62*, 589 (1979)
1.16 H.P. Baltes, R. Muri, F.K. Kneubuhl: "Spectral Densities of Cavity Resonances and Black Body Radiation Standards in the Submillimeter Wave Region", in *Submillimeter Waves*, Microwave Research Institute Symposia Series, Vol.XX, ed. by J. Fox (Polytechnic Press, Brooklyn 1971)
1.17 H.P. Baltes, E.R. Hilf: *Spectra of Finite Systems* (Bibliographisches Institut, Mannheim, Wien, Zürich 1976)
1.18 U. Kreibig: J. Phys. F*4*, 999-1014 (1974)
1.19 J. Phys. (Paris)*38*, Collo. C-2 (1977)
1.20 A.J. Kelly: J. Appl. Phys. *49*, 2621-2628 (1978)
1.21 J.C. Blake, A.J. Sabadell (eds.): *Proceedings of the First International Symposium on Coal-Oil Mixture Combustion* (Mitre Corporation, Mitrek Division, McLean, Va. 1978)
1.22 R.C. Flagan, S.K. Friedlander: "Particle Formation in Pulverized Coal Combustion—A Review", in *Recent Developments in Aerosol Sciences*, ed. by D.J. Shaw (Wiley, New York 1978)
1.23 P.J. Mayo, F.J. Weinberg: Proc. R. Soc. London A *319*, 351-371 (197o)
1.24 S. Chippett, W.A. Gray: Combust. Flame *31*, 149-159 (1978)
1.25 W.J. Reid: *External Corrosion and Deposits: Boilers and Gas Turbines* (American Elsevier, New York 1971)
1.26 W. Strauss: *Industrial Gas Cleaning* (Pergamon Press, Oxford 1976)
1.27 M. Sittig: *Particulates and Fine Dust Removal Processes and Equipment* (Noyes Data Corporation, Park Ridge, N.J. 1977)
1.28 J.C. Slater, K.H. Johnson: Phys. Today *27*, 34-41 (1974)
1.29 G.A. Ozin: Acc. Chem. Res. *10*, 21-26 (1977)
1.30 J.H. Sinfelt: Acc. Chem. Res. *10*, 15-20 (1977)
1.31 W.J. Walton (ed.): *Inhaled Particles and Vapors, IV* (Pergamon Press, Oxford 1977)
1.32 N. Nelson, J.L. Whittenberg (eds.): *Human Health and the Environment—Some Research Needs*, U.S. Department of Health, Education and Welfare publication no. NIH 77-1277
1.33 K.N. Palmer: *Dust Explosions and Fires* (Chapman and Hall, London 1973)
1.34 S. Twomey: *Atmospheric Aerosols* (Elsevier, Amsterdam, Oxford, New York 1977)
1.35 H. Israel: *Atmospheric Electricity*, Vol.I, transl. by D.B. Vaakov, B. Benny (National Technical Information Service, Springfield, Va.,TT67-513941/1, 1971)
1.36 H. Israel: *Atmospheric Electricity*, Vol.II, transl. by D.B. Yaakov, B. Benny (National Technical Information Service, Springfield, Va.,TT67-513941/2, 1973)
1.37 Y.S. Sedunov: *Physics of Drop Formation in the Atmosphere*, transl. by D. Lederman (Wiley, New York 1974)
1.38 H.R. Pruppacher, J.D. Klett: *Microphysics of Clouds and Precipitation* (D. Reidel, Dordrecht 1978)
1.39 S.H. Schneider, W.W. Kellog: "The Chemical Basis for Climate Change" in *Chemistry of the Lower Atmosphere*, ed. by S.I. Rasool (Plenum Press, New York, London 1973)
1.40 Harshvardhan, R.D. Cess: Tellus *XXVIII*, 1-10 (1976)
1.41 J.E. Hansen, W.-C. Wang, A.A. Lacis: Science *199*, 1065-1068 (1978)
1.42 S.I. Rasool (ed.): *Chemistry in the Lower Atmosphere* (Plenum Press, New York, London 1973)
1.43 *Topics in Interstellar Matter* (D. Reidel, Dordrecht 1977)
1.44 C.G. Granqvist, G.A. Niklasson: J. Appl. Phys. *49*, 3512-3520 (1978)
1.45 R. Gabler: *Electrical Interaction in Molecular Biophysics: An Introduction* (Academic Press, New York 1978)

1.46 V.A. Parsegian: "Long-Range Forces in the Biological Milieu" in *Annual Review of Biophysics and Bioengineering*, ed. by L.J. Mullins, W.A. Higgins, L. Stryer (Annual Review Inc., Palo Alto 1973)
1.47 J.N. Israelachivili: Quart. Rev. Biophys. *6*, 341-387 (1974)
1.48 W.H. Marlow, N. Abuaf: "Particle Sampling from High Temperature and Pressure Gas Streams", *Proceedings of the 1978 Symposium on Instrumentation and Control for Fossil Demonstration Plants*, ANL-78-62. Available from NTIS (Distribution Category UC-89)
1.49 D.A. Lundgren, F.S. Harris, Jr., W.H.Marlow, M. Lippmann, W.E. Clark, M.D. Durham: *Aerosol Measurement* (University Presses of Florida, Gainesville 1979)
1.50 B.Y.H. Liu (ed.): *Fine Particles: Aerosol Generation, Measurement, Sampling, and Analysis* (Academic Press, New York 1976)
1.51 E.D. Hinkley (ed.): *Laser Monitoring of the Atmosphere*, Topics in Applied Physics, Vol.14 (Springer, Berlin, Heidelberg, New York 1976)

2. The Kinetics of Ultrafine Particles

J. R. Brock

With 6 Figures

The subject of ultrafine particles (ufp)[1] is perhaps currently the most challenging and interesting area in aerosol science. This subject entails the difficult area of nucleation processes to explain the origin of most ufp. Analysis of the evolution in size, composition, space and time of ufp involves one with current problems in statistical mechanics, kinetic theory, probability theory, quantum chemistry, etc. One also encounters very difficult measurement problems for ufp, although these will not be discussed here.

Ultrafine particles also play important roles in such diverse areas as atmospheric chemistry and air pollution [2.1], catalysis [2.2], combustion [2.3], and gas dynamics [2.4], to name just a few.

In view of the many diverse aspects of the study of ufp, this survey, even though limited to consideration of the kinetics, requires additional limitation. The approach adopted is to emphasize exact, or at least nonempirical methods in the description of ufp kinetics. Experimental data and empirical formulas are employed only to substantiate or to illustrate particular theoretical approaches. A fairly recent and excellent survey exists already [2.5] on experiment and theory for ufp. Several extensive expositions including ufp are also available [2.6,7]. Some specific topics are omitted altogether. These include particle deposition and filtration, dynamics of charged particles, and a general review of ufp in turbulent flows.

The first section gives a brief summary of some of the significant parameters in the study of the dynamics of ufp. The second section presents a survey of methods for description of the evolution of a ufp aerosol in size/composition, space, and time. It begins with consideration of the Brownian-particle approximation and its description in statistical physics. This is followed by a survey of the nucleation and coagulation processes. The third and final section deals with the description of transfer processes to single particles. Problems arising in specifying boundary conditions (accommodation coefficients) are covered. Two examples are discussed.

---

[1]Ultrafine particles are rather arbitrarily defined here as particles with equivalent radius in the submicrometer range.

These are the isothermal drag force and the thermal force on a spherical particle. Various difficulties in current theory for these phenomena are cited.

2.1 Aerosol System Parameters

The essential features of this discussion are available in HIDY and BROCK [2.6]. Consequently, only a brief review will be given.

Aerosol particles in nature and many technical applications are very complex in terms of their morphology, internal structure, chemical composition, etc. As a consequence, little progress has been made in dealing with this complexity in terms of theories of evolution of aerosols. A very large set of parameters is necessary to describe an aerosol system.

Many current theories for aerosol systems begin with the assumption that the particles are rigid spheres. This permits some reduction in the complexity. Additional simplification is also made by assuming that the gas is monatomic or that for specific systems a polyatomic gas may be treated as a monatomic gas.

In this idealized model, some of the important dimensionless parameters frequently appearing in the study of aerosols are:

Knudsen Number $$Kn_i = \lambda_g/R_i \tag{2.1}$$

Mach Number $$Ma_i = |\underline{V}_i|/\bar{v}_g \tag{2.2}$$

Brown Number $$Br_i = \bar{v}_i/\bar{v}_g \tag{2.3}$$

Schmidt Number $$Sc_i = R_i^2\, n_g \lambda_g \tag{2.4}$$

where, the mean free path, λ_g, of host gas molecules is defined in terms of the viscosity coefficient, μ_g, the number density n_g, the molecular mass m_g, and the mean thermal speed $\bar{v}_g$ of the gas:

$$\mu_g = 0.499\, n_g m_g \bar{v}_g \lambda_g \quad . \tag{2.5}$$

The subscript "i" represents a property of the particles of radius R_i. $\underline{V}_i$ is the mean velocity of particles relative to the host gas. $\bar{v}_i$ is the mean thermal speed of particles. HIDY and BROCK [2.6] give a detailed discussion of these and other parameters of importance for ultrafine particles.

An ordering of other length and time scales is necessary for a complete description of aerosol system dynamics. In addition properties of the particles such as internal energy, viscosity, etc., may be important in some applications. These various quantities will be introduced as they arise in the following sections.

2.2 Evolution of an Aerosol of Ultrafine Particles

In the aerosol literature, one sees frequently macroscopic balance equations for the evolution of an aerosol in space, time, and particle size or composition. As an example, the evolution of the singlet density function, $n_i(\underline{r},t)$ at position $\underline{r}$ and time t for the aerosol phase can be written

$$\partial n_i(\underline{r},t)/\partial t + \nabla \cdot n_i\underline{q} + \nabla \cdot n_i\underline{V}_i = \sigma_{ij\alpha} \quad , \quad i,j = 1,2,3,\ldots \tag{2.6}$$

where n_i is the number concentration of particles with i momomers, $\underline{q}$ is the mean fluid velocity, $\underline{V}_i$ is the mean velocity of i-particles relative to the mean fluid velocity and $\sigma_{ij\alpha}$ indicates the rate of change of n_i due to coagulation, and evaporation or condensation of molecular species α, which obeys the conservation equation

$$\partial n_\alpha(\underline{r},t)/\partial t + \nabla \cdot n_\alpha\underline{q} + \nabla \cdot n_\alpha\underline{V}_\alpha = \sigma_{\alpha i} \quad , \quad i = 1,2,3,\ldots \tag{2.7}$$

where $\sigma_{\alpha i}$ denotes the rate of change for molecular species α of concentration n_α due to condensation and evaporation processes involving the aerosol phase. Equations (2.6,7) are easily generalized for an arbitrary number of molecular species.

There is also a continuity equation for the host gas, usually considered as inert relative to the processes of (2.6,7). In addition, complete macroscopic specification would require the conservation equations of energy and momentum, which would be coupled to (2.6,7). The effects of turbulence can be introduced through stochastic definitions for the macroscopic variables of the system. This approach has been used in describing "dusty gas" flows [2.8] and in analysis of gas dynamics of expansion flows with condensation [2.4].

However, the various macroscopic conservation equations used to describe such systems are incomplete. Most obviously, the various coefficients in the conservation equations are at this level phenomenological. Also, these equations give no information on statistical fluctuations which may be of importance.

2.2.1 General Theories of Evolution of ufp Aerosol Systems

The purpose of this section is to outline developments in the theory of the evolution of aerosol systems in terms of three levels of description:

1) Macroscopic theory (phenomenological description).
2) Mesoscopic theory (master equations).
3) Microscopic theory (molecular description).

This section opened with an example of the macroscopic theory which is based, of course, on the conservation laws. The "mesoscopic" description (a term due to VAN KAMPEN [2.9]) permits knowledge not only of the average behavior of an aerosol but also of its stochastic behavior through so-called master equations. However, this mesoscopic level of description may require (in complex systems) some physical assumptions as to the transition probabilities between states describing the system. Finally, the microscopic approach attempts to develop the theory of an aerosol from "first principles" – that is, through study of the dynamics of molecular motion in a suitable phase space. Master equations and macroscopic theory appear from the microscopic theory by the reduction of the complete dynamical description of the system in a suitable phase space to small subsets of chosen variables.

In terms of these three levels of description, we examine the following aspects of the evolution of a ufp aerosol system:

1) Brownian-particle approximation.
2) Nucleation regime.
3) Particle-growth regime.

The Brownian-Particle Approximation

For the Brownian-particle approximation, it is assumed that the aerosol system is stable over some experimental time. Particles are stable and their motions are uncorrelated. A special case in which the particles are treated as rigid bodies is frequently termed the Rayleigh gas. Of course, a characteristic feature of an aerosol is its inherent instability due to coagulation and deposition. Therefore, this approximation is limited to experimental times $T_{ex} \lll \tau_{ijc}$ (time of collision for all i,j) and the system is unbounded.

As emphasized by HIDY and BROCK [2.6] an important feature of an aerosol is the random motion of the particles known as Brownian motion. The study of Brownian motion has a long history and has played a key role in the development of modern statistical physics.

The traditional theory deals with the motion of a particle in a fluid at equilibrium. The fluid supplies a stochastic force as displayed in the Langevin equation [2.10]:

$$m_i(d\underline{q}_i/dt) = -\beta_i\underline{V}_i + \underline{A}_i(t) \tag{2.8}$$

where β_i is the drag coefficient of particle i with mass m_i and $\underline{A}_i(t)$ is the stochastic force acting on the particle. With assumptions as to the statistical character of $\underline{A}_i(t)$, one can derive from this macroscopic balance equation the equation describing the diffusion current of Brownian particles.

From the diffusion equation, the random character of Brownian motion is shown explicitly by the "uncertainty principle" [2.11] for times long compared to the particle relaxation time:

$$\Delta q_i \Delta r_i \geqq D_i \tag{2.9}$$

where Δq_i and Δr_i are, respectively, the dispersions in particle velocity and position and D_i is the Brownian diffusion coefficient of particle i.

The phenomenological, stochastic theory of Brownian motion exemplified by (2.8) has been superseded by microscopic [2.12] and "mesoscopic" [2.9,13] approaches.

The aim of the microscopic approach has been to obtain exact theories of Brownian motion from "first principles" which permit generalization to the study of many diverse statistical phenomena. As an example, one can begin the microscopic theory by writing down the classical Hamiltonian H for a system containing N_p Brownian particles and N_g (monatomic) gas molecules:

$$\begin{aligned} H = &\sum_{j=1}^{N_p} \frac{P_j^2}{2M} + \sum_{1 \leqq j < k \leqq N_p} \omega(|\underline{R}_j - \underline{R}_k|) + \sum_{\alpha=1}^{N} \frac{p_\alpha^2}{2m} \\ &+ \sum_{1 \leqq \alpha < \beta \leqq N_g} \mu(|\underline{r}_\alpha - \underline{r}_\beta|) + \sum_{j=1}^{N_p} \sum_{\alpha=1}^{N_g} \phi(|\underline{R}_j - \underline{r}_\alpha|) \end{aligned} \tag{2.10}$$

where lower case letters refer to gas molecules, capital letters refer to particles, and ω, μ, and ϕ represent the indicated potential energies of interaction. ϕ accounts for the particle–gas interaction whose nature will be discussed later. Momenta are indicated by $\underline{P}_i$ and $\underline{p}_\alpha$.

The Liouville equation is, therefore

$$\begin{aligned} &\frac{\partial f}{\partial t}(\underline{x}_1, \ldots, \underline{x}_{N_p}; \underline{x}_1, \ldots, \underline{x}_{N_g}; t) \\ &= \mathscr{L} f(\underline{x}_1, \ldots, \underline{x}_{N_p}; \underline{x}_1, \ldots, \underline{x}_{N_g}; t) \end{aligned} \tag{2.11}$$

where

$$\begin{aligned} \underline{x}_j &= \{\underline{r}_j, \underline{P}_j\} \\ \underline{x}_\alpha &= \{\underline{r}_\alpha, \underline{p}_\alpha\} \end{aligned} \tag{2.12}$$

and

$$\mathscr{L} = -\sum_{j=1}^{N_p} \frac{\underline{P}_j}{M} \cdot \frac{\partial}{\partial \underline{r}_j} - \sum_{\alpha=1}^{N_g} \frac{\underline{p}_\alpha}{m} \cdot \frac{\partial}{\partial \underline{r}_\alpha}$$

$$+ \sum_{1 \leq j < k \leq N_p} \sum \frac{\partial \omega_{jk}}{\partial \underline{r}_j} \cdot \left(\frac{\partial}{\partial \underline{P}_j} - \frac{\partial}{\partial \underline{P}_k}\right)$$

$$+ \sum_{1 \leq \alpha < \beta \leq N_g} \sum \frac{\partial \mu_{\alpha\beta}}{\partial \underline{r}_\alpha} \cdot \left(\frac{\partial}{\partial \underline{p}_\alpha} - \frac{\partial}{\underline{p}_\beta}\right) \tag{2.13}$$

$$+ \sum_{j=1}^{N_p} \sum_{\alpha=1}^{N_g} \frac{\partial \phi_{j\alpha}}{\partial \underline{r}_j} \cdot \left(\frac{\partial}{\partial \underline{P}_j} - \frac{\partial}{\partial \underline{p}_\alpha}\right) \quad .$$

Generalization to include nonspherical particles and molecules is straightforward.

Several techniques have been used to derive from (2.11), or generalization thereof, generalized Fokker-Planck and master equations as well as Langevin and transport equations [2.12,14,15].

In the Brownian-particle approximation, collisions between particles do not occur, and the potential $\omega(|\underline{R}_j - \underline{R}_k|)$ does not contribute to the evolution of the system which is governed by the evolution of the host gas. Also, the potentials μ and ϕ with interaction distances ℓ_{ig} and ℓ_g are (usually) assumed to be short range — that is, $\ell_{ig} \ll \lambda_{ig}$, $\ell_g \ll \lambda_g$. λ_{ig} is the mean free path for particle-gas collisions.

The mesoscopic approach begins with a master equation for the probability density $\mathscr{P}(\tilde{\Gamma},t)$ where $\mathscr{P}(\tilde{\Gamma},t)d\tilde{\Gamma}$ is the joint probability that at time t in a suitable phase space the Brownian particles are in the range $\tilde{\Gamma}_1 + d\tilde{\Gamma}_1, \tilde{\Gamma}_2' + d\tilde{\Gamma}_2, \ldots$. For a Markov process, $\mathscr{P}(\tilde{\Gamma},t)$ obeys the master equation

$$\frac{d}{dt} \mathscr{P}(\tilde{\Gamma},t) = \int \left[\mathscr{W}(\tilde{\Gamma}|\tilde{\Gamma}')\mathscr{P}(\tilde{\Gamma}',t) - \mathscr{W}(\tilde{\Gamma}'|\tilde{\Gamma})\mathscr{P}(\tilde{\Gamma},t)\right] d\tilde{\Gamma}' \tag{2.14}$$

where the kernel $\mathscr{W}(\tilde{\Gamma}|\tilde{\Gamma}')$ for $\tilde{\Gamma} \neq \tilde{\Gamma}'$ is the transition probability per unit of time from $\tilde{\Gamma}'$ to $\tilde{\Gamma}$ in phase space. From equations such as (2.14) the Fokker-Planck equation for the Brownian particle in an equilibrium gas can be derived [2.10].

The assumption in Brownian motion is always $m_i \gg m_g$. It is possible to determine the range of validity of the Brownian-particle assumption (and the range of validity of the Fokker-Planck equation) for the model of particle and gas as rigid spheres — the Rayleigh gas. For homogeneous host gas a Fokker-Planck equation can be written and solved exactly; for the same problem, a transport or master equation can be studied with an exact scattering kernel [2.16].

The basic spatially homogeneous master equation for the relaxation of Brownian-particle energy $x \equiv m_i V_i^2/2kT$ in a host gas at temperature T is

$$\frac{\partial \mathscr{P}(x,t)}{\partial t} = \int_0^\infty K(y,x)\mathscr{P}(y,t)dy - Z(x)\mathscr{P}(x,t) \tag{2.15}$$

where

$$Z(x) = \int_0^\infty K(x,y)dy \quad .$$

$\mathscr{P}(x,t)$ is the particle energy distribution function and $K(x,y)$ is the probability per unit time of a collision scattering the particle with energy x to an energy in the range y,dy.

The initial value problem of (2.15) describes the time evolution of $\mathscr{P}(x,t)$ from some prescribed $\mathscr{P}(x,0)$ to the final equilibrium Maxwellian

$$\mathscr{P}(x,\infty) = 2\pi^{-\frac{1}{2}} x^{\frac{1}{2}} e^{-x} \quad .$$

A dimensionless time scale τ is introduced in terms of the particle-gas collision time τ_{igc}

$$\tau = \frac{8}{3}\gamma Z(0)t$$

where

$$\gamma = m_g/m_i$$

and

$$Z(0) = 8\pi R_i^2 n_g \left(\frac{kT}{m_i}\right)^{\frac{1}{2}} \gamma^{-\frac{1}{2}}$$

is the collision frequency (approximately τ_{igc}^{-1}).

For the rigid-sphere kernel, for which $K(x,y)$ and $Z(x)$ are known explicitly, the solution of (2.15) involves the eigenvalue problem

$$\int_0^\infty B(x,y)N(y)dy = -\varepsilon N(x) \tag{2.16}$$

where

$$B(x,y) = (x/y)^{\frac{1}{2}} e^{(\frac{1}{2})(x-y)} K(x,y) - Z(x)\delta(x - y) \quad . \tag{2.17}$$

For $\tau \gg \tau_{igc}$ (2.15) has the approximate solution

$$\mathscr{P}(x,t) \simeq 2\pi^{-\frac{1}{2}}x^{\frac{1}{2}}e^{-x} + 2^{\frac{1}{2}}\pi^{-\frac{1}{2}}e^{-x/2} \cdot \sum_{\varepsilon=1}^{\infty} a_\varepsilon N_\varepsilon(x)e^{-\lambda_\varepsilon \tau} \quad . \tag{2.18}$$

N_ε and λ_ε are the eigenfunctions and discrete eigenvalues. There exist also continuum modes, when the restriction $\tau > \tau_{igc}$ is not introduced.

In the limit $\gamma \to 0$, from (2.15) the corresponding Fokker-Planck equation can be derived [2.17]

$$\frac{\partial \mathscr{P}}{\partial \tau} = \hat{F}\mathscr{P}(x,\tau) = \left[x \frac{\partial^2}{\partial x^2} + (x + \frac{1}{2}) \frac{\partial}{\partial x} + 1\right]\mathscr{P}(x,\tau) \quad . \tag{2.19}$$

$\hat{F}$ is self-adjoint and the eigenvalue condition leads to the complete orthonormal set of solutions:

$$\tilde{\phi}_\varepsilon(x) = \left[\varepsilon!\Gamma(\varepsilon + \frac{3}{2})\right]^{\frac{1}{2}} x^{\frac{1}{4}} e^{-x/2} L_\varepsilon^{\frac{1}{2}}(x) \quad . \tag{2.20}$$

These are the Laguerre functions of order 1/2 with eigenvalues

$$\lambda_\varepsilon = \varepsilon \ ; \quad \varepsilon = 0, 1, 2, \ldots \infty \quad . \tag{2.21}$$

Therefore, in this approximation:

$$\mathscr{P}(x,\tau) \simeq 2\pi^{-\frac{1}{2}}x^{\frac{1}{2}}e^{-x} + 2^{\frac{1}{2}}\pi^{-\frac{1}{2}}x^{\frac{1}{4}}e^{-(\frac{1}{2})x} \cdot \sum_{\varepsilon=1}^{\infty} a_\varepsilon \tilde{\phi}_\varepsilon e^{-\varepsilon\tau} \tag{2.22}$$

where

$$a_\varepsilon = \int_0^\infty \frac{\tilde{\phi}_\varepsilon(x)}{\tilde{\phi}_0(x)} \mathscr{P}(x,0)dx \quad . \tag{2.23}$$

Figure 2.1 shows the eigenvalues for the eigenvalue problem (2.16) as calculated by HOARE and KAPLINSKY [2.16] by a variational method. It is seen that as γ decreases, the eigenvalues approach the integer eigenvalues of the Fokker-Planck equation. The area on the left of Fig.2.1 corresponds to the continuum region ($\tau \sim \tau_{igc}$). An analogous correspondence exists for the eigenfunctions.

For this model, the Brownian-particle regime (as defined by the applicability of the Fokker-Planck equation) exists to good approximation for $(m_g/m_i) < 0.001$ with the limitation $\tau \gg \tau_{igc}$.

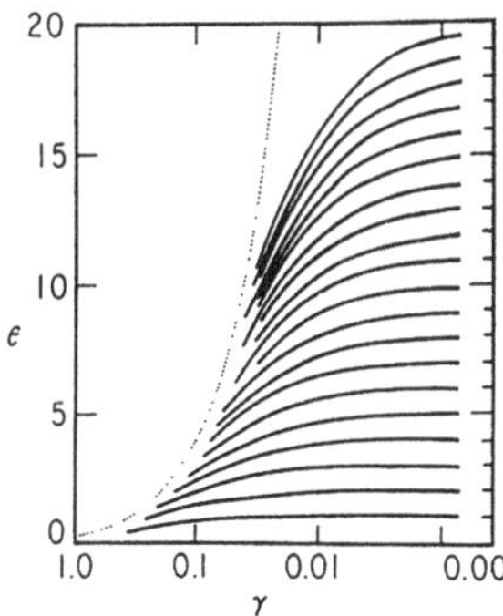

Fig.2.1. Eigenvalues of the energy scattering kernel for a hard-sphere Brownian test particle as a function of mass ratio, $\gamma = (m_g/m_i)$. [2.16]

Mesoscopic approaches based on master equations have been applied to other problems in aerosol physics including coagulating systems [2.18] and nucleating systems [2.19]. These areas will be discussed below.

In principle, generalized kinetic equations derived by microscopic analysis starting from (2.11) or suitable generalizations could be used to describe the evolution, including statistical and hydrodynamic fluctuations, of an arbitrary ufp aerosol system. However, such general treatment of an arbitrary aerosol system appears to be unlikely in the near future.

One may then ask what applicable theory exists for the nonequilibrium evolution of a ufp aerosol system in the Brownian-particle approximation. One has still the restriction $T_{ex} \ll \tau_{ijc}$ so that particle-particle collisions can be neglected.

If the aerosol particles can be regarded as large molecular components in a dilute gas system, then the evolution of the system can be studied with the well-developed kinetic theory of dilute gases. This requires in the Brownian-particle approximation that the particles doe not alter the state of the host gas and that changes in the state of the host gas over distances L_g are negligible over distances of the order of the particle size. If one introduces the following restrictions:

$$Ma_i \ll 1 \tag{2.24}$$

$$(R_g/\lambda_g) \;,\quad (R_g/L_g) \ll 1 \tag{2.25}$$

$$(n_i/n_g) \ll 1 \quad , \quad \text{for all } i \tag{2.26}$$

$$(R_i/L_g) \ll 1 \quad , \quad \text{for all } i \tag{2.27}$$

$$(R_i/\lambda_g) \ll 1 \quad , \quad \text{for all } i \tag{2.28}$$

$$T_{ex} \ll \tau_{ijc} \quad , \quad \text{for all } i \text{ and } j \tag{2.29}$$

then the (assumed spherical) particles can be regarded as components of a dilute monatomic gas mixture whose evolution according to (2.24-29) is described by the appropriate set of Boltzmann equations for the mixture, i.e.,

$$\frac{\partial f_i}{\partial t} + \underline{V}_i \cdot \nabla f_i = J(f_i f_g) \ , \ i = 1,2, \ \dots \tag{2.30}$$

$$\frac{\partial f_g}{\partial t} + \underline{V}_g \cdot \nabla f_g = J(f_g f_g) \ , \tag{2.31}$$

where $f_i(\underline{V}_i;\underline{r},t)$ is the velocity distribution function of i-particles and J is the Boltzmann collision operator. With the additional restriction that $(m_g/m_i)^{\frac{1}{2}} << 1$, (2.30,31) describe what has been termed a quasi-Lorentz gas [2.20]. MASON and co-workers, in a series of papers, have shown that "normal" solutions of (2.30,31) [2.21] give results for the mean particle motion, $\underline{V}_i$, in nonequilibrium suspending gases which agree, upon allowing for diffuse and inelastic scattering for gas-particle collisions, with those obtained by free-molecular momentum calculations [2.5,6] for single, spherical aerosol particles. This agreement also implies that results from the kinetic theory of polyatomic gases [2.22] could be applied both to polyatomic suspending gases and to [with generalizations of (2.25-29)] non-spherical particles with internal degrees of freedom, although this has not been done.

Similar agreement with free-molecular momentum calculations for spherical particles has been obtained through study of Brownian diffusion in nonequilibrium gases as described by the Fokker-Planck equation [2.23,24].

In the Brownian-particle approximation, therefore, one has an exact description of the nonequilibrium evolution of the aerosol system within the limitations imposed by restrictions (2.24-29). Such systems are partially described by (2.6) with $\sigma_{ij\alpha} = 0$, i.e.

$$\partial n_i(r,t)/\partial t + \nabla \cdot n_i \underline{q} + \nabla \cdot (n_i \underline{V}_i) = 0 \ . \tag{2.32}$$

Nucleation

The initial step in the evolution of most ultrafine particles (ufp) is their birth in the gas phase by a nucleation process. A number of distinct nucleation processes are recognized. A primary distinction exists between nucleation occurring on existing particles or ions (heterogeneous nucleation) and nucleation occurring in the absence of particles or ions (homogeneous nucleation). Homogeneous nucleation processes have been further subdivided as heteromolecular or homomolecular. The heteromolecular process occurs when particles are formed by the interaction of two or more chemical species, as in the H_2SO_4-H_2O system. Homomolecular nucleation involves only a single chemical species.

The majority of the current theories of nucleation correspond to the macroscopic approach discussed above, and consequently have the characteristic deficiencies noted. The original macroscopic theory of homogeneous homomolecular nucleation is attributed to BECKER and DÖRING [2.25] (the B - D or "classical" theory). Subsequent work on macroscopic theories of nucleation have proceeded from the B - D theory. There exist many extensive reviews of vapor-phase nucleation [2.6,26-30]. Consequently, only more recent developments in nonmacroscopic approaches will be cited. However, a brief discussion of the "classical" B - D theory is necessary to put more recent work into proper perspective.

BECKER et al. [2.31] have indicated that the B - D theory for homogeneous nucleation agrees with experiment to within 5 percent for more than a dozen chemical compounds (including ethanol, n-alkanes, water, n-alkylbenzene, ethane, and halogenated methane) and have shown that agreement within 7 percent holds for menthol in He carrier gas in the highest-temperature cloud-chamber study reported to date. This agreement refers to the ability of the B - D theory to predict the critical supersaturation ratio as a function of temperature. It is generally recognized that such prediction is not a critical test of the details of the nucleation process. This is emphasized by the fact that the original Lothe-Pound modification [2.32] of the B - D theory predicted nucleation rates some 10^{17} times larger than the B - D theory, and that experiments were unable to resolve this discrepancy!

A number of experimental studies [2.33-35] have sought more direct information on the nucleation process by carrying out experiments which attempt to determine some measure of the nucleation rate by measurement of size distributions or moments of such distributions. While these experiments provide additional information on the nucleation process, they still depend on phenomenological coefficients for their interpretation and cannot examine details of the nucleation process.

The microscopic approach to nucleation problems has apparently not yet been carried out. There have been a number of mesoscopic developments for homogeneous nucleation [2.19,36-38]. The mesoscopic approach is successful in giving information on fluctuations, which are, of course, central to the process of nucleation. In this, the mesoscopic approach improves on the macroscopic approach. However, the transition probability is not known from "first principles" and, therefore, must retain some phenomenological elements.

Recent developments in the application of molecular dynamics promise a resolution of many questions in nucleation theory. A review of simulation methods in nucleation theory and of their results is given in [2.39]. Consequently, a survey will not be undertaken here. However, some of the results of SCHIEVE and co-workers [2.40-42] will be discussed here as they throw light on several questions of importance in terms of modelling aerosol growth from vapor condensation.

ZUREK and SCHIEVE have carried out molecular-dynamics calculations for 3- and 2-dimensional systems with a square-well hard-core gas. Their results for a 2-dimensional system, which agree well with 3-dimensional calculations, are more extensive and will be discussed here. Several interesting conclusions emerge from their molecular-dynamics study.

One of the assumptions of all the "classical theories" is that clusters grow by monomer addition:

$$n_i + n_1 \rightarrow n_{i+1} \tag{2.33}$$

and that the process

$$n_i + n_j \rightarrow n_{i+j} \tag{2.34}$$

may be neglected. The reasoning is the intuitive one that, since $n_1 \gg n_i$, (2.33) must follow. The molecular-dynamics calculations of ZUREK and SCHIEVE indicate that the process (2.34) of coagulation of clusters may contribute importantly to the evolution of the system, as shown in Table 2.1. The distribution of clusters in adiabatic systems of varying total energy is shown in Fig.2.2. E^* is the reduced energy ($E/\tilde{\varepsilon}$) of the system (per particle per degree of freedom) consisting of 100 hard-core square-well disks of mass 6.628×10^{-23}g. The square-well parameters are $\sigma_1 = 2.98 \times 10^{-8}$cm, $\sigma_2 = 1.96\sigma_1$, $\tilde{\varepsilon} = 2.305 \times 10^{-14}$erg. T^* is the reduced temperature ($kT/\tilde{\varepsilon}$).

Table 2.1. Number of clusters used in formation and produced in destruction of larger clusters ($E^* = 0.3$). For $\ell < m$ element (ℓ,m) gives the number of clusters of size ℓ used in formation of clusters of size m, while the element (m,ℓ) gives the number of clusters of size ℓ produced in dissociation of clusters of size m

	1	2	3	4	5	6	7	8	9	10	11	12
1	0	446	530	272	122	45	12	11	10	12	7	1
2	450	0	530	134	57	32	3	4	2	3	4	0
3	523	523	0	272	57	10	5	4	1	1	2	0
4	252	152	252	0	122	32	5	2	3	1	2	0
5	127	60	60	127	0	45	3	4	3	2	1	0
6	48	21	20	21	48	0	12	4	1	1	1	2
7	10	4	2	2	4	10	0	11	2	1	2	0
8	7	3	8	2	8	3	7	0	10	3	2	0
9	10	2	1	2	2	1	2	10	0	12	4	0
10	14	4	2	2	2	2	2	4	14	0	7	0
11	10	1	1	1	2	2	1	1	1	10	0	1
12	0	2	0	0	0	0	0	0	0	2	0	0

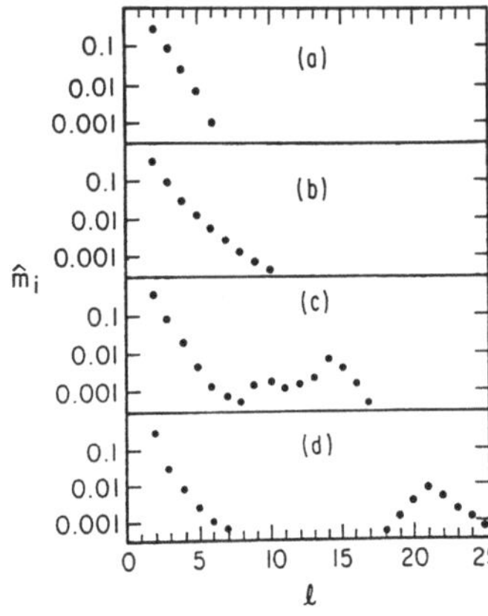

Fig.2.2a-d. Fraction of clusters, $\hat{m}_i$, as a function of cluster size, ℓ , in adiabatic systems of varying total energy, E^*. (a) $E^* = 0.5$, $T^* = 0.66$; (b) $E^* = 0.3$; $T^* = 0.56$; (c) $E^* = 0.0$, $T^* = 0.53$; (d) $E^* = -0.2$; $T^* = 0.51$. (c) and (d) show occurrence of supercritical clusters. [2.40]

As E is decreased one observes a change from the unimodal distribution for subcritical clusters to a bimodal form indicating growth of supercritical clusters. Because the system is adiabatic, the biomodal distributions also represent stationary states in which there are maximum supercritical cluster sizes, which, if exceeded, result in destruction of that supercritical cluster size; new bonds formed in the system increase the cluster kinetic energy and decrease the pressure of the monomer gas. In the future it would be desirable to extract from the molecular-dynamics calculation accurate values for the free energy of formation of clusters. Such calculations would resolve the differences between the B - D theory and the Lothe-Pound theory. In the future, molecular-dynamics calculations should make possible development of correct mesoscopic and microscopic theories of homogeneous and even heterogeneous nucleation.

Current theories of vapor-phase nucleation usually assume homogeneous systems. Nucleation occurs in many cases under highly inhomogeneous conditions, such as in shock waves, nozzle exhausts, flames, etc. The applicability of current theories under these conditions needs careful examination [2.43].

Coagulation

After formation of ultrafine particles (ufp) by nucleation, subsequent growth of supercritical clusters occurs by condensation and coagulation. Coagulation is discussed in this section dealing with evolution of aerosols of ufp. The discussion is limited to Brownian coagulation which is the principal mode of coagulation for ufp.

There are already a number of surveys of coagulation [2.5-7,44,45] which cover the derivation (mesoscopic and macroscopic) and important developments in analytical and numerical techniques for solving the homogeneous coagulation equation

$$\frac{\partial n_i}{\partial t} = \frac{1}{2} \sum_{j+k=i} b_{jk} n_j n_k - n_i \sum b_{ij} n_j \quad , \tag{2.35}$$

or its continuous version

$$\frac{\partial n(m)}{\partial t} = \frac{1}{2} \int_0^m dm' b(m - m', m') n(m - m') n(m') - n(m) \int_0^\infty dm' b(m, m') n(m') \quad . \tag{2.36}$$

More recently [2.46-48] general macroscopic equations have been used to examine the evolution of the continuous singlet density function, $n(m,\underline{r},t)$:

$$\begin{aligned} \frac{\partial n(m,\underline{r},t)}{\partial t} + \nabla \cdot \underline{q} n(m,\underline{r},t) &= \nabla \cdot \underline{\underline{K}} \cdot \nabla n(m,\underline{r},t) \\ &+ \frac{1}{2} \int_0^m dm' b(m - m', m') n(m - m', \underline{r}, t) n(m, \underline{r}, t) \\ &- n(m,\underline{r},t) \int dm' b(m', m) n(m', \underline{r}, t) \\ &- \frac{\partial}{\partial m} [\Psi(m) n(m,\underline{r},t)] + \underline{G} \cdot \nabla n(m,\underline{r},t) \\ &+ \sum_p \dot{\nu}_p + \sum_i \dot{\nu}_{N_i} \quad . \end{aligned} \tag{2.37}$$

$n(m,\underline{r},t)$ is the number of aerosol particles with mass m in the range m, dm at point $\underline{r}$ in space at time t. $\underline{q}$ is the fluid velocity. $\underline{\underline{K}}$ is a general diffusion tensor. The second and third terms on the right account for coagulation and the fourth for condensation. The fifth term accounts for forced or gravitational settling with velocity $\underline{G}$. $\dot{\nu}_p(m,\underline{r},t)$ is the rate of input of particles from discrete sources, and $\dot{\nu}_{N_i}(m,\underline{r},t)$ is the rate of production of particles by nucleation. In simulation studies of nucleating systems, (2.6) is coupled to macroscopic conservation equations of condensing monomers. Figures 2.3,4 show some of the rate processes and moments in an approximate analysis of the evolution of sulfuric acid vapor according to (2.6) (using the B - D theory) and the monomer conservation equation. Only Brownian coagulation is considered. As is evident from the figures, after about 5 min, sufficiently high concentrations of particles exist to suppress homogeneous nucleation. Subsequently, growth of particles occurs by condensation and coagulation as demonstrated in Figs.2.3,4.

Macroscopic equations have also been proposed [2.49] for the evolution by coagulation in composition "space" of density functions for particles composed of an arbitrary number of chemical species. The analytical solutions for such

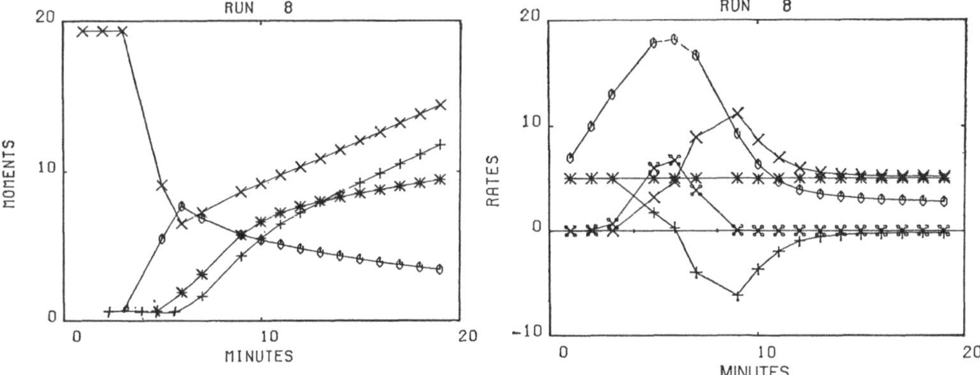

Fig.2.3. Moments of particle size distribution for nucleation and growth of sulfuric acid aerosol from sulfuric acid vapor at low humidity. Number concentration, cm^{-3}, $\times$ 2E5, 0; mass concentration, g/cc $\times$ 1E-13, + ; surface, cm^{-1} $\times$ 1E-7, * ; 1/Number, cm^{3} $\times$ 1E-5, X

Fig.2.4. Rate processes in evolution of sulfuric acid aerosol nucleated from sulfuric acid vapor at low humidity. Monomer concentration g/cc $\times$ 2E-14, 0 ; total rate g/cc-s 2E-16, + ; input rate, g/cc-s $\times$ 2E-16, * ; nucleation rate g/cc-s $\times$ 2E-18, %

multicomponent aerosols have also been investigated [2.50,51] when the coagulation coefficient has analytically tractable forms.

Macroscopic developments of the type discussed have the obvious difficulty that the expression b(m,m') for the coagulation coefficient is phenomenological. b(m,m') for Brownian coagulation has been studied experimentally [2.52-61]. An excellent survey of experimental methods and results for Brownian coagulation has been given by MERCER [2.62]. Not made clear in Mercer's survey is the fact that most of the coagulation studies cited above, except for the first three, have been made using the condensation nucleus counter as a means of measuring the total particle concentration. The condensation nucleus counter (CNC) is now known [2.63] to be inaccurate; there is a serious loss in counting efficiency as particle size decreases. Most of the studies cited used the CNC with particles less than the order of 0.2 μm diameter; therefore, the studies are in error and underestimte the particle concentration with the result that the calculated b(m,m') from experimental data is too large. In addition, it is clear that studies which follow only the decrease of total number concentration with time by sampling suffer from many other deficiencies such as wall losses, unknown effect of polydispersity of the aerosol, etc. All these experimental problems appear to have been overcome by WAGNER and KERKER [2.52] who follow the change with time of the particle size distribution using light-scattering techniques.

In general, a large set of parameters is necessary to characterize the coagulation process between two particles i and j. Considerable simplification results if the following assumptions are made: 1) particles are spherical, liquid droplets which coalesce upon collision; 2) host gas is at equilibrium; 3) particle temperatures are equilibrated with the gas; 4) particle-particle and gas-particle interactions are short range; 5) $n_i \ll n_g$.

Under these conditions, the two "rocks of faith" in the theory for b_{ij} are: 1) the free molecular limit where the particles behave as large molecules (Kn_i, $Kn_j \to \infty$) and 2) the mixed limit ($Kn_i \to 0$, $Kn_j \to \infty$) where i is a continuum sphere and j behaves like a large molecule in a dilute host gas with a mutual diffusion coefficient D_{jg}.

In case 1, b_{ij} is

$$b_{ij} = \left[8\pi kT(m_i + m_j)/m_i m_j\right]^{\frac{1}{2}}(R_i + R_j)^2 \tag{2.38}$$

which can be modified easily to account for various assumed particle interaction energies [2.6].

In case 2, b_{ij} is

$$b_{ij} = 4\pi(D_{ig} + D_{jg})(R_i^* + R_j^*) \tag{2.39}$$

where D_{ig} and D_{jg} are the mutual diffusion coefficients of i and j in host gas and R_i^* and R_j^* are the radii of the "spheres of influence" about the particles. It is assumed that at the distance $(R_i^* + R_j^*)$ coalescence occurs. Equation (2.39), therefore, contains still an assumption [2.10]. It can be made more precise if we suppose that, in addition, $m_i \ggg m_j$ so that

$$b_{ij} = 4\pi D_{jg} R_i \tag{2.40}$$

which, when multiplied by n_j is Maxwell's equation for the growth rate of a spherical drop with negligible vapor pressure.

It is assumed in (2.38,40) that the sticking (and coalescence) efficiency is unity. Deviations from unity are sometimes represented by introducing a coefficient α_c for this efficiency—that is, $b'_{ij} = \alpha_c b_{ij}$. However, there is no viable theory to support this assumption.

For practical application of macroscopic theory where Brownian motion is the only factor causing collision, the semi-empirical theory of FUCHS [2.7] is most widely used because it gives correctly the values of b_{ij} in the limits, (2.38,40). The careful experimental results of WAGNER and KERKER [2.52] for b_{ij} at large Kn_i agree with the theory of FUCHS to within 8 to 10 percent upon the assumption of a coalescence efficiency of unity.

LOYALKA [2.64] has given values of b_{ij} based on his calculations of the condensation rates of spheres in noncontinuum regimes; his values differ by around 10 percent from values calculated from the equation of FUCHS. b_{ij} has also been calculated [2.65,66] for the transition regime of Kn_i using the steady state

Fokker-Planck equation [2.10]. As SITARSKI and SEINFELD point out, their theory does not give the correct limit values (2.38,40).

Current theories of coagulation do not account explicitly for an additional Knudsen number which enters into the coagulation process. As two particles with finite Kn_i approach during collision, a new Knudsen number becomes important: $Kn_{ijc} = \lambda_g/\Delta_{ij}$, where Δ_{ij} is the interparticle distance. For approach of two particles of the same size, Kn_{ijc} varies over the limits $(\lambda_g n_i^{1/3}, \infty)$. For atmospheric pressure and normal particle concentrations, $\lambda_g n_i^{1/3} \ll 1$ and represents the Knudsen number for the average interparticle spacing. As two particles approach, their Knudsen and hydrodynamic boundary layers must interact; this interaction would be characterized by Kn_{ijc}. Hydrodynamic and other interaction effects are well known to be important in Brownian particle systems in liquids [2.67-69].

An additional test of theories for b_{ij} under the ideal conditions listed above would be afforded by variations for a colliding pair of $m_i, m_j, R_i, R_j, m_g, \lambda_g$. However, since variations of these parameters are accounted for in the two limiting regimes, one might expect that Fuch's equation, which interpolates between the two regimes, would still give a good estimate. As has been noted [2.6], simple interpolation methods (such as "Sherman's universal relation") represent transition-region behavior surprisingly well for a wide variety of transport phenomena.

Outside the ideal equilibrium conditions discussed, the number of parameters necessary to specify the coagulation process is large. STEPANOV [2.10] has attempted to develop a theory of coagulation in nonequilibrium systems starting with an equation of continuity for the N-particle distribution function $f_N(\underline{r}_1, \ldots, \underline{r}_N, x_1, \ldots, x_N; t)$ where $\underline{r}_i$ is the position vector for the ith particle, with mass x_i:

$$\frac{\partial f_N}{\partial t} + \sum_{i=1}^{N} \frac{\partial}{\partial \underline{r}_i} \cdot \underline{r}_i f_N + \sum_{i=1}^{N} \frac{\partial}{\partial x_i} \dot{x}_i f_N = 0 \quad . \tag{2.41}$$

$\dot{x}_i$ is an assumed growth law for the ith particle.

This theory, however, does not appear to be complete in that the particle growth laws are phenomenological and, therefore, avoid the difficult questions posed above. One must still await a general theory for coagulation in nonequilibrium gases.

Sticking efficiency and coalescence upon collision are two factors that produce considerable uncertainty in the analysis of Brownian coagulation of ufp aerosols. As has been noted there is no viable theory for sticking efficiency. Coalescence may be assumed for liquid ufp, but the coagulation of solid ufp must generally lead to the formation of aggregates with highly variable morphology.

For aggregates or flocs which may arise on Brownian coagulation of solid ufp, a number of simulation techniques have been proposed [2.71-78]. These techniques involve computer simulation of coagulation based on either the simple (constant coagulation coefficient) Smoluchowski theory [2.10] or improvements on the theory by Muller [2.79] to allow for the effect of polydispersity in calculating the collision rate. Some simple empirical relations have been obtained from experimental work supported by the simulation studies quoted. For the density of aggregates, BEECKMANS [2.77] proposed:

$$(\rho/\rho_0) = (m_0/m)^h \tag{2.42}$$

where ρ and m are respectively the density and mass of the aggregate. ρ_0 and m_0 are the values for primary particles making up a cluster. h is a constant which is zero for liquids and has a maximum of 0.64 for solid spherical particles. A relation has also been proposed for the projected area to total primary-particle number by MEDALIA and HECKMAN [2.76]:

$$N = (SA/A_p)^a \tag{2.43}$$

N is the number of particles, SA is the projected aggregate surface area (as from electromicrographs) and A_p is the projected area of a single particle. These relations (2.43,44), of course, are based on empirical observation of a relatively small number of aggregates and may have limited validity.

STÖBER [2.80,81] and KOPS et al. [2.82] have found an empirical relation between the aerodynamic diameter, D_a, of aggregates and the number of particles, N, in the aggregate. KOPS et al. [2.82] report

$$D_a \propto N^{\varepsilon}(\rho/\rho_0)^{\frac{1}{2}} D_{1\,geo} \exp(C_\ell \ln^2 \sigma_{1\,geo}) \tag{2.44}$$

where $\rho_0 = 1$ g/cc, $D_{1\,geo}$ is the geometric mean diameter and $\sigma_{1\,geo}$ the geometric standard deviation of the primary-particle size and ρ is the density of the aerosol material. C_ℓ is a constant. The exponent $\delta = 1/6$ for approximately linear aggregates and $\delta = 1/3$ for more or less spherical aggregates. These results are based only on two aerosol materials – iron oxide and gold. Just as for (2.42,43), (2.44) is based on limited empirical observation and may, therefore, have limited validity.

Obviously, the analysis of coagulation of solid aggregates of ufp represents an interesting area for additional study.

2.3 Transfer Processes to Single Particles

As pointed out earlier in Sect.2.2 outside the free molecular regime, there is no general microscopic kinetic equation for description of the evolution of ufp aerosols in nonequilibrium host gas. There are macroscopic equations such as (2.6) or the generalized Langevin equation [2.6]:

$$\frac{d\underline{q}_i}{dt} = - \beta_i(\underline{q}_i - \underline{q}) + \frac{\underline{F}_{F_i}}{m_i} + \frac{\underline{F}_{NE_i}}{m_i} + \frac{\underline{\mathscr{F}}_i}{m_i} \tag{2.45}$$

for motion of particle i in the absence of interparticle collisions. β_i is the particle drag coefficient. $\underline{F}_{F_i}$ and $\underline{F}_{NE_i}$, respectively, are the forces associated with external fields and various gaseous nonuniformities. $\underline{\mathscr{F}}_i$, of course, is the stochastic force associated with fluctuations in the host gas.

Equation (2.45) is a macroscopic equation whose application would require knowledge of the various forces and of the nature of the fluctuations in the nonequilibrium host gas. This section is concerned with partial description of the evolution of a ufp aerosol in terms of (2.45) or (2.6) by study of the dynamics of single spherical, rigid particles. Much of the earlier work in this area is covered in several references [2.5,6,83].

2.3.1 Transfer Calculations and Accommodation Coefficients

In nonequilibrium systems one can postulate the separation between stochastic and mean values, although it may be very difficult to arrive at explicit descriptions. The situation is simplified somewhat by the introduction of several restrictions. In the discussion which follows, it is assumed that the Brownian-particle approximation holds. Also the additional restriction of quasistationarity is introduced for the transfer processes. A single spherical particle in this regime is characterized by the four parameters, Kn_i, Ma_i, Sc_i, and Br_i. In addition, other parameters arise such as the accommodation coefficients specifying the transfer efficiencies between host gas and particle, and particle properties, including the thermal conductivity k_i and viscosity μ_i.

Transfer Calculations

With these restrictions, the mean forces on a particle are calculated by integrating the momentum flux over the particle surface

$$\underline{F}_i = - \int dS_i \underline{n} \cdot \left(\int_+ dv_g m_g \underline{v}_g \underline{v}_g f_g^+ + \int_- dv_g m_g \underline{v}_g \underline{v}_g f_g^-\right) \quad . \tag{2.46}$$

f_g^+ and f_g^- indicate velocity distribution functions of the gas at the particle surface, defined for molecules for which $\underline{v}_g \cdot \underline{n} > 0$ and $\underline{v}_g \cdot \underline{n} < 0$, respectively. $\underline{n}$ is a unit vector outwardly normal to the surface element dS.

The other transfer processes can be calculated by

$$\Phi_i^{(\alpha)} = \int dS_i \underline{n} \cdot \left(\int_+ dv_g^{(\alpha)} \underline{v}_g^{(\alpha)} f_g^{(\alpha)} + \int_- dv_g^{(\alpha)} \underline{v}_g^{(\alpha)} f_g^{(\alpha)} \right) \tag{2.47}$$

for the total rate of molecular transfer of chemical species α to the surface. The total rate of heat transfer, H_i, for a monatomic gas is

$$H_i = \int dS_i \underline{n} \cdot \left(\int_+ dv_g \frac{1}{2} m_g v_g^2 \underline{v}_g f_g^+ + \int_- dv_g \frac{1}{2} m_g v_g^2 \underline{v}_g f_g^- \right) \quad . \tag{2.48}$$

Equations (2.46-48) can be generalized to account for nonspherical particles and polyatomic gases.

The Boltzmann equation must be solved with appropriate boundary conditions to obtain f_g^+ and f_g^-. The full Boltzmann equation has not been solved analytically or numerically. Current approximate methods for extracting the desired information from the Boltzmann equation are covered in detail in a recent reference [2.84]. In view of this review, discussion of these methods will not be given. It is sufficient to indicate some of the principal methods which have been employed. These are moment or integral methods for specific molecular scattering laws, the use of "models" (of which the BGK model is the simplest) for the collisions term $J(f_g f_g)$, and direct simulation by Monte Carlo or molecular-dynamics techniques. All these methods have been applied to problems of interest in the study of ufp.

Accommodation Coefficients

No general theory exists for determining f_g^+. One is faced with the difficult, practical problem of the scattering of molecules from the surface of a particle with (possibly) various unknown impurities and absorbed surface layers. More than this, it is well known that very small particles or clusters have properties which can differ substantially from those of the bulk material. Such deviations have been known for some time [2.6] and within the past few years have received increased attention [2.85]. There are not only the effects of curvature on the thermodynamic equilibrium between a particle and its host gas, but also for the particle changes in crystal structure, electronic properties, optical characteristics, dynamical properties of crystal lattices, etc. There are extensive reviews on gas-surface interactions [2.86]. However, almost all work done to date has concerned experiments or theory for the surfaces of bulk material usually under vacuum, degassed conditions. As noted above, such conditions have little to do with ufp.

One can, in principle, define the problem of surface boundary conditions by the expression

$$\underline{n} \cdot \underline{v}_g f_g^+(\underline{v}_g) = \int_- |\underline{n} \cdot \underline{v}_g'| P(\underline{v}_g' \rightarrow \underline{v}_g) f_g^-(\underline{v}_g') dv_g' \tag{2.49}$$

where $P(\underline{v}_g' \cdot \underline{v}_g)dv_g$ gives the probability that a gas molecule hitting the surface with velocity $\underline{v}_{g1}'$ would be reflected with a velocity lying in the range $\underline{v}_g, dv_g$. A still more ambitious program would be to treat the gas-surface problem by starting with the Liouville equation for the entire system. Neither of these approaches have led yet to practical results, although a number of models have been proposed for determination of f_g^+ [2.84].

In practice, the accommodation coefficient, α, proposed by MAXWELL [2.87] has been applied most often to the problem of determining f_g^+:

$$f_g^+(\underline{v}_g) = \alpha\, f_g^{(0)+}(\underline{v}_g) + (1 - \alpha) f_g^-(\underline{v}_g') \tag{2.50}$$

where

$$f_g^{(0)+} = n_g^+(m_g/2\pi kT^+)^{3/2} \exp(-m_g v_g^2/2kT^+)$$

and (2.51)

$$\underline{v}_g' = \underline{v}_g - 2\underline{n}(\underline{n} \cdot \underline{v}_g) \quad .$$

In this simple assumption, a fraction α of the molecules leave the surface as a Maxwellian stream equilibrated with the surface, and a fraction $(1 - \alpha)$ are specularly reflected. In terms of the scattering kernel of (2.49), Maxwell's boundary condition is:

$$P(\underline{v}_g' \rightarrow \underline{v}_g) = \alpha f^{(0)+} + (1 - \alpha)\delta(\underline{v}_g' - \underline{v}_g) \quad . \tag{2.52}$$

This model is immediately suspect—a fast molecule striking the surface should have a smaller probability of sticking than a slower molecule. Also one would expect a dependence of sticking probability on angle of incidence. Improvements on (2.50) along these lines have been proposed but have not led too far.

Accommodation coefficients are often defined through the fluxes

$$\alpha_j = (\phi_j^- - \phi_j^+)/(\phi_j^- - \phi_{0j}^+) \tag{2.53}$$

where $\phi_j^\pm$, j = 1,2,3,4,5 are incident and reflected fluxes of momentum components, kinetic energy, and mass respectively, and ϕ_{0j}^+ refers to perfect accommodation. The α_j have been termed the Knudsen accommodation coefficients [2.88].

Transfer processes in the free-molecular regime can be expressed exactly in terms of these Knudsen accommodation coefficients [2.88]. However, this is not the case in the slip-flow regime as has been demonstrated by Kuscer. The situation in the transition regime has not been investigated yet.

One approach has been suggested by WALDMANN [2.89] and WALDMANN and VESTNER [2.90]. This involves the use of a truncated version of Maxwell's moment equations. Boundary conditions for Waldmann's "generalized hydrodynamics" are established by the methods of nonequilibrium thermodynamics with restrictions arising from the second law and Onsager-Casimir symmetries. This gives, of course, phenomenological coefficients for the boundary laws. These coefficients are implicitly dependent on the accommodation coefficients, but the nature of this dependence cannot be determined solely from the phenomenological theory.

Nonphenomenological theories have been proposed [2.88,91,92] starting with (2.49) or its equivalent. Kuscer has shown that in the slip-flow regime a suitable set of accommodation coefficients is sufficient. This approach is summarized below.

For the half-space problem where a gas fills the space $x \geqq 0$ bounded by a plane wall, one can define a set of dynamic quantities $Q_j = Q_j(|v^*_{gx}|, v^*_{gy}, v^*_{gz})$, where $\underline{v}^*_g = (m_g/2kT_s)^{\frac{1}{2}}\underline{v}_g$ is the dimensionless molecular velocity and T_s is the surface temperature. In Kuscer's analysis, the velocity distribution of molecules incident upon a surface is assumed to have the form of a slightly distorted Maxwellian, $f_k \propto \exp(-\underline{v}^{*2})(1 + \varepsilon Q_k)$. The accommodation coefficient α_{jk} for the quantity Q_j with incident distribution f_k is defined by

$$\alpha_{jk} = (\phi^-_{jk} - \phi^+_{jk})/(\phi^-_{jk} - \phi^+_{0jk}) \quad . \tag{2.54}$$

ϕ^-_{jk} and ϕ^+_{jk} are respectively the incident and reflected fluxes of Q_j and ϕ^+_{0jk} is the reflected flux of Q_j if the surface were perfectly accommodating.

With the introduction of the average with respect to P:

$$\tilde{F}(\underline{v}^*_g) = \int_- P(-\underline{v}^*_g \rightarrow -\underline{v}^{*\prime}_g)F(\underline{v}^{*\prime}_g)dv^{*\prime}_g \quad , \tag{2.55}$$

and of the Maxwellian average over the velocity half-space, $v_x > 0$:

$$\langle F(\underline{v}^*_g)\rangle = 2\pi^{-3/2} \int_+ F(\underline{v}^*_g)\exp(-\underline{v}^{*2}_g)dv^*_g \quad . \tag{2.56}$$

α_{jk} can be written as

$$\alpha_{jk} = \left\langle v^*_{gx}Q_j(\underline{v}^*_g)[Q_k(\underline{v}^*_g) - \tilde{Q}_k(\underline{v}^*_g)]\right\rangle / \left\langle v^*_{gx}Q_j(\underline{v}^*_g)\left\{Q_k(\underline{v}^*_g) - \pi^{\frac{1}{2}}\langle v^*_{gx}Q_k(\underline{v}^*_g)\rangle\right\}\right\rangle \quad . \tag{2.57}$$

For the velocity and temperature slip problems the following set of dynamic quantities are used

$$Q_1 = |v^*_{gx}| \quad , \quad Q_2 = v^*_{gy} \quad , \quad Q_3 = v^*_{gz} \quad , \quad Q_4 = v^{*2}_g \quad , \quad Q_5 = |v^*_{gx}|v^*_{gy} \quad ,$$

$$Q_6 = |v^*_{gx}|v^*_{gz} \quad , \quad Q_7 = |v^*_{gx}|v^{*2}_g \quad .$$

This leads to nine accommodation coefficients: α_{11}, $\alpha_{22} = \alpha_{33}$, α_{14}, α_{44}, $\alpha_{25} = \alpha_{36}$, $\alpha_{55} = \alpha_{66}$, α_{17}, α_{47}, and α_{77}. α_{11}, $\alpha_{22} = \alpha_{33}$, α_{44}, and α_{14} are respectively the Knudsen accommodation coefficients of normal momentum, tangential momentum, energy, and thermal creep. The additional coefficients α_{25}, α_{55}, α_{17}, α_{47}, and α_{77} are termed "second-order" accommodation coefficients and do not have any simple physical meaning. Also they do not appear to be directly measurable.

More generally, KUSCER points out that owing to the self-adjointness of the surface scattering operator, which can be defined from (2.49), the array α_{jk} is symmetric, $\alpha_{jk} = \alpha_{kj}$ which may be regarded as an example of Onsager's law for conjugate transport phenomena. For perfect accommodation, all $\alpha_{jk} = 1$, and for specular reflection all $\alpha_{jk} = 0$.

For various assumed models of surface-gas interactions it is possible to establish the correspondence between the α_{jk} and the Knudsen accommodation coefficients, α_j. However, it is clear that these various models do not deal with the complexity inherent in surface-gas interactions in the slip regime.

In summary, transfer processes in the free-molecular regime can be expressed exactly in terms of the α_j, (2.53). The slip and possibly the transition regimes require the introduction of much more complex accommodation coefficients, including "second-order" coefficients which do not appear to be directly measurable. Of course, even the α_j are not known a priori for ultrafine particles. Unfortunately, the accommodation coefficients must be regarded now as adjustable parameters in the theory for ultrafine particles.

2.3.2 Momentum Transfer

The summation of the momentum flux over the surface of a spherical particle by (2.46) gives for nonequilibrium host gas the force acting on the particle. The exact distribution functions f^+_g and f^-_g are to be obtained from the stationary solution of the Boltzmann equation

$$\underline{v}_g \cdot \nabla f_g = J(f_g f_g) \quad , \tag{2.58}$$

in the vicinity of the particle using suitable boundary conditions. The use of the Boltzmann equation implies, of course, that the force calculated according to (2.46) will be deterministic. As pointed out in Sect.2.2, (2.58) yields no information on the stochastic forces associated with Brownian motion. Whether or not the forces calculated by (2.46) for particles in the transition regime correspond to the mean forces defined in (2.45) will depend upon the nature of the nonuniformity in the host gas.

Throughout this discussion the restriction $Ma_i \ll 1$ is applied. This is reasonable for many ultrafine-particle systems. Relaxation of this restriction is possible, but this leads to difficult problems outside the free-molecular regime.

For nonequilibrium host gas, a number of particle forces, $\underline{F}_{NE_i}$ arise. These include: isothermal drag force, thermal force, photophoretic force, diffusion force, stress force, and other additional cross effects in combined flows of heat, mass, and momentum.

Of these various forces, only the isothermal drag force and the thermal force will be discussed here. The isothermal drag force presents perhaps the simplest of the nonequilibrium, noncontinuum phenomena. Yet, as will be shown, the current state of knowledge of the drag force is inadequate. The thermal force provides an example of the complexity inherent in particle motion in nonequilibrium host gas.

Isothermal Drag Force

Numerous experimental and theoretical investigations have been made of the isothermal drag force on a particle in the noncontinuum regime. Much of the work is summarized in several references [2.5-7]. As will be evident in this discussion, there exists even now no satisfactory comprehensive description of this phenomenon.

Exact Results

Exact results are known for the isothermal drag force in the two limits $Kn_i \to 0$ and $Kn_i \to \infty$ for $Ma_i \ll 1$. For $Kn_i \to 0$, Stokes' law is readily derived from the Navier-Stokes equation with stick boundary conditions

$$\underline{F}_i = 6\pi\mu_g R_i \underline{V}_i \tag{2.59}$$

where μ_g is the viscosity of the host gas, and $\underline{V}_i$ the relative velocity of the spherical particle. For slip boundary conditions, Basset's result [2.93] is well known

$$\underline{F}_i = 6\pi\mu_g R_i \underline{V}_i \left(\frac{1+2C_m Kn_i}{1+3C_m Kn_i}\right) \tag{2.60}$$

where C_m is the velocity slip coefficient. For perfect slip in the continuum limit, the following expression is obtained [2.93]

$$\underline{F}_i = 4\pi\mu_g R_i \underline{V}_i \quad . \tag{2.61}$$

Exact results are known also in the free-molecular regime, $Kn_i \to \infty$ for $Ma_i \ll 1$ [2.94]

$$\underline{F}_i^* = (8/3)R_i^2 n_g (2\pi m_g kT)^{1/2}(1 + \pi\alpha/8)\underline{V}_i \qquad (2.62)$$

where α is Maxwell's accommodation coefficient, (2.50) and n_g, m_g are the number density and mass of host gas molecules. If Knudsen's accommodation coefficients, (2.53), are applied one obtains [2.95]

$$\underline{F}_i^* = (\pi/3)[(2 - \alpha_n + \alpha_t)(4/\pi) + \alpha_n]R_i^2 n_g (2\pi m_g kT)^{1/2}\underline{V}_i \qquad (2.63)$$

where α_n and α_t are respectively the normal and the tangential momentum accommodation coefficients. Equation (2.63) reduces to (2.62) for $\alpha_n = \alpha_t$. Unfortunately, very little is known about α_n and its relation to α_t.

It is only in the two limits, $Kn_i \to 0$, $Kn_i \to \infty$ for $Ma_i \ll 1$ that exact results are available. One may conclude that a necessary, but not sufficient condition for a satisfactory noncontinuum description of the isothermal drag force is its reduction to (2.59,61,62) or (2.63) in the appropriate limits.

Approximate Descriptions

The standard of comparison for the isothermal drag force on a spherical particle for $Ma_i \ll 1$ has for many years been the empirical relation of MILLIKAN and coworkers [2.96] based on extensive experimental measurements:

$$\underline{F}_i = 6\pi R_i \underline{V}_i \{1 + Kn_i[A + B\exp(-CKn_i^{-1})]\} \qquad (2.64)$$

where A, B, C are empirical constants which depend on the nature of the host gas and particle. Table 2.2 presents some representative values of these constants.

Examination of (2.64) in the limit $Kn_i \to \infty$, shows that

$$A + B = 2.25/(1 + \pi\alpha/8) \qquad (2.65)$$

from (2.62). For small Kn_i, the term first order in Kn_i in (2.64) shows that

$$A = C_m \quad , \qquad (2.66)$$

the velocity slip coefficient.

Millikan's empirical equation (2.64) was derived from experimental data for gas-particle systems with nearly perfect accommodation ($\alpha = 0.8 - 1.0$). Therefore, (2.64) is not likely to be correct as to its dependence on gas-particle accommodation. This is seen easily from the fact that the constant, B, calculated from (2.65,66) takes on zero and negative values: $B \leqq 0$, when Maxwell's accommodation

Table 2.2. Representative experimental values of constants in Millikan's equation (2.64)

Particle	Gas	A	B	C	Ref.
oil	air	1.246	0.42	0.87	[2.96]
M-300 silicon oil	A	1.49 (3%)	0.35 (10%)	1.0 (20%)	[2.97]
PH-300 silicon oil	A	1.42 (3%)	0.50 (10%)	0.9 (20%)	[2.97]
PN-200 silicon oil	A	1.46 (3%)	0.40 (10%)	0.8 (20%)	[2.97]
Paraffin	A	1.28 (3%)	0.55 (10%)	0.8 (20%)	[2.97]
M-300	N_2	1.45 (3%)	0.40 (10%)	0.9 (20%)	[2.97]
M-300	CO_2	1.48 (3%)	0.40 (10%)	1.1 (20%)	[2.97]
M-300	H_2	1.53 (3%)	0.40 (10%)	1.6 (20%)	[2.97]
NaCl (spherical)	A	1.20 (3%)	-	-	[2.98]

coefficient, α, is less than the order of 0.73. This is the case regardless of whether the classical value of C_m or more accurate values such as (2.67) are used. Such a decrease in B is neither observed experimentally [2.6] nor is it in accord with theoretical results. In general, a critical test of empirical or theoretical results for the dynamics of ultrafine particles is afforded by examination of the behavior of the result with respect to the surface-gas accommodation.

The velocity slip coefficient has been studied by several investigators using Maxwell's boundary conditions [2.84,99]. LOYALKA et al. [2.99] give the following expression for C_m derived from the BGK model

$$C_m = [(2 - a)/\alpha](\pi^{\frac{1}{2}}/2)(1 + 0.1621\alpha) \quad . \tag{2.67}$$

This is close to "exact" values calculated numerically from the BGK analysis.

It is known that Maxwell's and Knudsen's boundary conditions are not correct in the slip regime. Kuscer's analysis [2.88] of the velocity slip problem using the linearized Boltzmann equation yields:

$$C_m = [(2 - \alpha_{25})/\alpha_{22}](\mu_g^{(1)}/\mu_g)^2\left(1 + (2/\pi)\alpha_{22}[(2 - \alpha_{55})/(2 - \alpha_{25})] - \alpha_{25}/2\right) \quad . \tag{2.68}$$

Here α_{22} is the tangential momentum accommodation coefficient, α_{25} and α_{55} are "second-order" coefficients. $\mu_g^{(1)}$ is the Chapman-Enskog first-order approximation to the viscosity coefficient. Equation (2.68) is, of course, valid for general gas-surface scattering kernels (2.49). Equation (2.68) reduces approximately to

(2.67) with the assumption of Maxwell's gas-surface scattering kernel for Maxwellian molecules.

Integral Methods

Integral methods include all those which attempt solution of the moment equation (Maxwell's equation of transfer) of the Boltzmann equation. The well known integral methods which have been applied include Mott-Smith's bimodal distribution [2.100], Grad's 13 moment equations [2.101], Lees' two-stream Maxwellian [2.102], and Waldmann's higher-order hydrodynamics and boundary conditions [2.103].

The attraction of the integral method is its ease of application to practical problems to obtain approximate descriptions over the complete range of Kn_i, $0 < Kn_i < \infty$.

An important objection to integral methods lies in the arbitrary truncation and closure of the moment equation. These procedures have never been justified analytically or numerically. A discussion of these points may be found in Lees [2.104] and KOGAN [2.95]. The limitations of the integral method will become apparent in this discussion.

GOLDBERG [2.105] used the 13 moment equations with Grad's boundary conditions to obtain the following expression for the isothermal drag force on a spherical particle:

$$\underline{F}_i = 30\pi\mu_g R_i \underline{V}_i \left\{\frac{2\pi+19\pi c_1 Kn_i+(12+30\pi c_1^2)Kn_i^2}{10\pi+105\pi c_1 Kn_i+(72+225\pi c_1^2)Kn_i^2+162c_1 Kn_i^3}\right\} \tag{2.69}$$

where $c_1 = (2-\alpha)/\alpha$. This equation gives correctly (2.59,61) in the limit $Kn_i \to 0$, but predicts a force in the free-molecular limit, $Kn_i \to \infty$, approximately six times too large.

LIU and SUGIMURA [2.106] applied Lees' method with $\alpha = 1$, to obtain

$$\underline{F}_i = 40\pi(8+\pi)\mu_g R_i \underline{V}_i (41 + 120\ Kn_i) \quad , \tag{2.70}$$

which agrees with (2.62) for $Kn_i \to \infty$, but is 80 percent high compared to Stokes' formula, (2.59).

PHILLIPS [2.107] also used Lees' method, but chose a different set of parametric functions from LIU and SUGIMURA. In addition Phillips used Maxwell's boundary conditions with the result

$$\underline{F}_i = 6\pi\mu_g R_i \underline{V}_i \left\{\frac{15-3c_1 Kn_i+c_2(8+\pi\alpha)(c_1^2+2)Kn_i^2}{15+12c_1 Kn_i+9(c_1^2+1)Kn_i^2+18c_2(c_1^2+2)Kn_i^3}\right\} \tag{2.71}$$

where $c_2 = 1/(2 - \alpha)$. This equation is correct in the limit $Kn_i \to \infty$, but fails in the limit $Kn_i \to 0$ to give (2.61). Also (2.71) obviously does not exhibit the required nonanalytic form in the free- near-free-molecular regime [see (2.79)] and is in error by over-30 percent (see Table 2.3) in its value of the near-free-molecular coefficient, p_1 [see (2.76) for definition of p_1]. Therefore, (2.71) is inexact and appears to be incorrect with regard to its dependence on the accommodation coefficient, α. However, for $\alpha \sim 1$, it agrees well ($\sim$ 2 percent) with Millikan's empirical equation.

Table 2.3. Experimental and calculated values of first-order near free molecular coefficient

Analysis	p_1
MILLIKAN's experiment [2.96]	0.38 (α = 0.895)
SCHMITT (average for oil drops in argon) [2.97]	0.33 (α = 0.82)
WILLIS' theory-BGK [2.112]	0.366 (α = 1.000)
KAN's theory-BGK [2.111]	0.380 (α = 1.000)
LIU and SUGIMURA theory-moment [2.106]	0.293 (α = 1.000)
PHILLIPS' theory-moment [2.107]	0.549 (α = 1.000) 0.503 (α = 0.895)
SUR's interpolation (2.80)	0.607 (α = 1.000)

Most of the previous developments have proceeded from Maxwell's boundary conditions which are known now to be in error. WALDMANN and co-workers have derived thermodynamically consistent boundary conditions for higher-order constitutive equations. For monatomic gases and Maxwell molecules, these constitutive equations reduce to Grad's 13 moment equations. However, WALDMANN has pointed out that Grad's boundary conditions are thermodynamically inconsistent. For the drag force problem, VESTNER and WALDMANN derived

$$\underline{F}_i = 6\pi\mu_g R_i \underline{V}_i \left(1 + \sum_{q=1}^{4} a_q Kn_i^q\right) \Bigg/ \left(1 + \sum_{q=1}^{5} b_q Kn_i^q\right) \tag{2.72}$$

where the coefficients a_q and b_q are functions of the velocity slip coefficient C_m, the temperature-jump coefficient C_t, the thermal-creep coefficient C_{cr}, a surface heat-conduction coefficient C_h, and the particle to gas thermal-conductivity ratio $\Lambda = k_i/k_g$. The first two coefficients are

$$\begin{aligned} a_1 &= 2C_t\Lambda/(\Lambda + 2) + 2C_m + 2C_h/(\Lambda + 2) \\ b_1 &= 2C_t\Lambda/(\Lambda + 2) + 3C_m + 2C_h/(\Lambda + 2) \end{aligned} \quad . \tag{2.73}$$

Waldmann's analysis makes clear that for the "isothermal" drag-force problem the shear motion of the gas produces a heat flow within the so-called Knudsen layer of the order of one mean-free-path thickness surrounding the particle. Kuscer's analysis by the linearized Boltzmann equation also demonstrates this. Of course, Kuscer's analysis penetrates more deeply in that it provides explicit values for the coefficients C_t, C_m, and C_h in terms of the generalized accommodation coefficients, α_{jk}. In Waldmann's analysis the former are simply phenomenological coefficients. One would expect that Waldmann's procedure would be valid in the slip regime because the derived boundary conditions are obtained by calculation of the interfacial entropy production in the "Knudsen layer". This is acceptable so long as $\lambda_g/R_i \ll 1$, so that the "Knudsen layer" can be regarded as an interfacial region. Indeed, (2.72) agrees in the limit $Kn_i \to 0$ with both (2.59,61). However, (2.72) is in error for $Kn_i \to \infty$ where (2.72) becomes

$$\underline{F}_i = (8/3)R_i^2 n_g(\pi^2 m_g kT_0)^{\frac{1}{2}} \underline{V}_i (C_m - \xi C_{cr}^2/C_h)^{-1} \tag{2.74}$$

where: $\xi = k_g \mu_g T_0/p_0^2 \lambda_g^2$, and subscript 0 indicates mean value. For $C_m \to \infty$, $F_i \to 0$ according to (2.74), which cannot be correct [see (2.62)]. For $Kn_i \to \infty$, (2.74) does agree with (2.62), and for perfect accommodation differs from (2.62) by around 30 percent. From this, however, one should not infer that (2.72) is necessarily valid outside the slip regime. Hydrodynamic theories, such as the one by which (2.72) is derived, represent some arbitrary truncation and closure of Maxwell's moment equation. These procedures have not been justified, and as a consequence, approximate agreement outside the slip regime may be only fortuitous. In any event, (2.72) does not display the established nonanalytic properties for large Kn_i [see (2.79)], it displays a dependence on Λ of the near-free-molecular coefficient p_1 [see (2.76)], which has not been observed experimentally, and it yields a value for p_1 for perfect accommodation in monatomic gas-particle systems which is around one-third of the corresponding experimental value (see Table 2.3).

Linearized Boltzmann Equation and Models

The principal difficulty in solving the Boltzmann equation lies in the analytically intractable collision term. For small disturbances from equilibrium, the collision term may be linearized. Another approach is the calculation of transfer processes about a particle using a relaxation model for the collision term. It would be expected that such models would be most successful in near-free-molecular conditions where the "free-streaming" terms are much more important than collisions between host-gas molecules. The so-called BGK model is perhaps the most widely applied of these models [2.5,6].

For the BGK model, (2.58) becomes

$$v_g \cdot \nabla f_g = \tilde{\delta}(f_g^{(0)} - f_g) \quad , \tag{2.75}$$

$f_g^{(0)}$ is a local Maxwellian and $\tilde{\delta}$ is a constant which be chosen to obtain correctly either the viscosity or thermal conductivity from (2.75).

Cercignani's results for isothermal drag based on solution of (2.75) by a variational method agree with Millikan's empirical equation to within 2 percent. The defect in these results is that they are purely numerical and assume perfect accommodation. SONE and AOKI [2.108] have studied isothermal drag by means of an expansion about $Kn_i = 0$ with the the linearized BGK model. Their results fit (2.64) well for small Kn_i.

An additional test of the various analyses of the drag-force problem is afforded by the near-free-molecular behavior. To first order in Kn_i^{-1}

$$F_i/\underline{F}_i^* = 1 - p_1 Kn^{-1} \tag{2.76}$$

where $\underline{F}_i^*$ is the free-molecular drag force.

From Millikan's empirical equation (2.64):

$$p_1 = (1 - BC)/(A + B) \quad . \tag{2.77}$$

Table 2.3 shows values of p_1 defined by (2.76) for various analyses of the drag-force problem. As might be expected, the BGK model gives values of p_1 closest to the experimental value. Of course, these results are only valid for near-perfect accommodation.

By use of an inner-and-outer expansion procedure for solution of the linearized Boltzmann equation. PAO and WILLIS [2.109] have shown for three dimensional problems that the expansion for any transfer function, such as F_i, in Kn_i^{-1} about $Kn_i^{-1} = 0$ is nonanalytic and has the form

$$Y_i/Y_i^* = 1 + y_1 Kn^{-1} + y_2 Kn^{-2} \ln(Kn_i^{-1}) + y_3 Kn_i^{-2} + \dots \quad . \tag{2.78}$$

The same nonanalytic expansion was also found by DORFMAN et al. [2.110]. KAN [2.111] has evaluated the first two constants for the drag-force problem:

$$F_i/F_i^* = 1 - 0.38\ Kn_i^{-1} + 0.0527\ Kn_i^{-2} \ln(Kn_i^{-1}) + \dots \quad . \tag{2.79}$$

Nothing yet is known about the convergence of the expansion [2.110]. The close relation of (2.78) to the density expansion of the transport coefficients has also been pointed out. If the numerical values in (2.79) are approximately correct, it is easy to see that the logarithmic behavior cannot be easily deduced from experimental results.

Interpolation

SHERMAN [2.113] has proposed what is sometimes termed "Sherman's universal relation" (SUR) which has been applied to a large variety of noncontinuum phenomena. In terms of this discussion, SUR has the form

$$1/F_i = 1/F_i^* + 1/F_{ic} \tag{2.80}$$

where F_{ic} is the continuum force, (2.59).

Actually, the form of (2.80) has been known for some time in mathematical analysis and engineering as an efficient means for interpolating between two limit values. It is not surprising that it succeeds, at least approximately, as shown in Fig.2.5 where (for $\alpha \sim 1$) SUR is compared with Millikan's empirical equation, (2.64). SUR agrees within about 5 percent with Millikan's equation over $0 < Kn_i < \infty$.

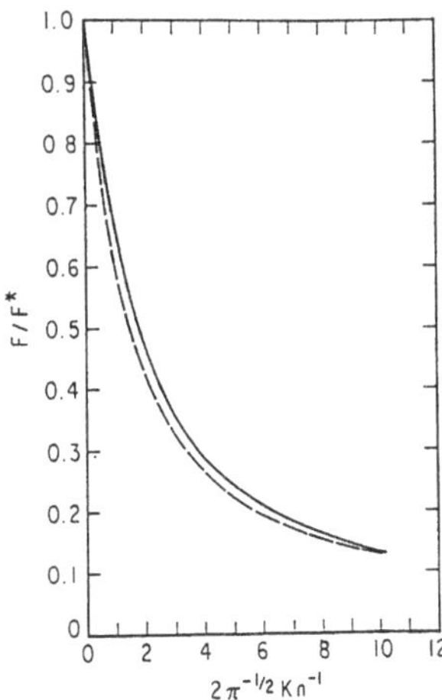

Fig.2.5. Isothermal drag force on spherical particle as a function of Knudsen number for the particle. Millikan's experimental data (——) from [2.96]. "SUR" interpolation formula (-----), see (2.80)

McCOY and CHA [2.114] have given explicit expressions for a number of transfer processes with various geometries using SUR.

A general method related to SUR has been proposed by MASON and co-workers for various noncontinuum phenomena. Using this method for the present problem, ANNIS et al. [2.115] give a formula for the drag force derived through the assumption

$$\underline{V}_i = \underline{V}_{i\,visc} + \underline{V}_{i\,diff} \tag{2.81}$$

where $\underline{V}_{i\,visc}$ is obtained from (2.59) and $\underline{V}_{i\,diff}$ is obtained from the diffusive flow in a binary mixture composed of the particles and the host gas. The final equation of ANNIS et al. is

$$\underline{F}_i = 6\pi\mu_g R_i \underline{V}_i / \{1 + u_1 Kn_i[(1 + u_2 Kn_i)/(1 + u_3 Kn_i)]\} \quad . \tag{2.82}$$

This equation contains three parameters u_1, u_2, and u_3. When these are determined from Millikan's constants A, B, C, a good fit of Millikan's data is obtained. When these parameters are calculated for elastic diffuse scattering, there is substantial error. Indeed, one does not recover the free-molecular limiting value.

Discussion

From this discussion of the drag force a number of tests have emerged for an acceptable theory. First, the theory must reduce to the exact results of (2.59-63). Next, it must contain the nonanalytic properties in the near-free-molecular regime, (2.79). Finally, gas-surface interactions must be in terms of general accommodation coefficients such as those introduced by KUSCER.

Even if all these tests were to be successfully passed, there remain some problems. Most obviously, the accommodation coefficients are not known for ultrafine particles. In addition, the role of additional parameters such as $Kn_L = \lambda_g/L$, where L is a characteristic distance in a finite system, remains to be investigated.

With the considerable difficulties still remaining in the analysis of the isothermal drag force for spherical ultrafine particles, it is not surprising that the analysis of this problem for nonspherical particles is incomplete. This subject is reviewed by FUCHS [2.7]. Recent kinetic-theory analyses of drag on nonspherical bodies in the free-molecular regime can be found in CERCIGNANI [2.84]. The excellent series of experimental studies by STÖBER and co-workers [2.80,81,116] as well as more recent work on this problem [2.82,117-119] may be consulted for further information.

Thermal Force

It has been known for some time [2.120] that a temperature gradient imposed on a gas causes suspended particles to migrate in the direction of decreasing temperature. The force causing this migration is termed the thermal force. There exist a number of reviews of this subject [2.5-7,121,122].

The thermal force is sufficiently complex that experimental investigations have often obtained widely different results. Recent advances in experiment and theory may resolve some of these differences. However, the theory of the thermal force still lacks a general rigorous development.

A much larger set of basic parameters is required for description of the thermal force than is necessary for the drag force. Transfer of heat to a particle by the host gas is central to the phenomenon. In the case of a polyatomic host gas, one must work from the more difficult kinetic theory of polyatomic gases [2.123]. Heat transfer at the particle-gas interface presents the problem of specifying the energy accommodation coefficients which often differ substantially from perfect accommodation. Additional complexity will appear in the discussion which follows.

The physical system to be described consists of a particle of radius R_i suspended in the middle of a host gas bounded (for convenience) by infinite parallel plates separated by a distance L. This physical configuration leads to the two Knudsen numbers: $Kn_i = \lambda_g/R_i$ and $Kn_L = \lambda_g/L$. For more complex geometry, additional boundary Knudsen numbers like Kn_L would arise. In addition, one must also consider a Mach number: $Ma_{it} = |\underline{V}_i \cdot \underline{e}|/\bar{v}_g$. Here, $\underline{e}$ is a unit vector in the direction of the temperature gradient. Also, the Brown number must be considered.

This survey begins with a summary of exact results for the thermal-force problem. Then approximate developments will be discussed in the light of available data. Finally, some developments for the thermal force in dense gases and liquids as well as new effects in dilute host gas will be described.

Exact Results

Exact results for the thermal force for a highly viscous, stable particle in dilute, monatomic host gas are available in the limits $Kn_i \to \infty$, $Kn_L \to 0$. The particle is in the free-molecular regime and the host gas may be treated as a Chapman-Enskog dilute gas in which the Knudsen layers next to the parallel plates have no influence on the state of the particle.

These limits permit one to attain exact theoretical developments by two different methods. In one of these [2.124,125] (2.58) is used with Maxwell's boundary conditions to give for the free-molecular thermal force:

$$\underline{F}_{it}^{*} = -(32/15)(k_{tr}\, R_i^2/\bar{v}_g)\nabla T \quad (2.83)$$

where k_{tr} is the translational part of the thermal conductivity (this is to extend the result approximately to polyatomic gases). In these limits, the particle can also be regarded as a large molecular component of the host gas (quasi-Lorentz gas) whose migration under a temperature gradient is described by the Chapman-Enskog theory for the thermal-diffusion effect. When only elastic collisions are considered, there is disagreement with (2.83); however, an accounting of inelastic collisions [2.126] yields results which imply (2.83).

Equation (2.83) agrees with experiment to within 5 percent for monatomic gases [2.97,98,127]. For polyatomic gases there is a more serious discrepancy (~ 20 percent) which may be due to the approximate nature of the kinetic theory of polyatomic gases employed as well as to the use of Maxwell's boundary conditions. Equation (2.83) also predicts an independence of the thermal force from the properties of the particle—a result also confirmed by the data of SCHMITT and JACOBSEN and BROCK.

By use of Knudsen's accommodation coefficients (2.53) for this problem in the same limits, the following result is found for the free-molecular thermal force [2.128]:

$$\underline{F}^{*}_{it} = -(32/15)R_i^2[1 + (\alpha_t - \alpha_n)/2](k_{tr}/\bar{v}_g)\nabla T \tag{2.84}$$

which reduces to (2.83) for $\alpha_t = \alpha_n$. Unfortunately, little is known about the relation between α_t and α_n so that in (2.84) they must be regarded for now as adjustable parameters.

One may conclude from the somewhat limited experimental data available that (2.83), with the restrictions noted earlier, is confirmed for monatomic host gases both as to the magnitude of the thermal force and its dependence on the physical properties of the particle. Additional confirmation is also provided by the implicit agreement of (2.83) with thermal-diffusion theory for a quasi-Lorentz gas.

Exact results are also available for the limits $Kn_i \to \infty$, $Kn_L \to \infty$ so that there are no intermolecular collisions in the host gas between the parallel plates. The only collisions are those of host-gas molecules with the particle and plates. With the important restriction that heating of the particle by thermal radiation can be neglected, the following result holds in this case [2.129]:

$$F_{it} = -(3/4)\pi R_i^2 p_C[(T_H/T_C)^{\frac{1}{2}} - 1] \tag{2.85}$$

where perfect accommodation is assumed. T_H and T_C are the temperatures of the hot and cold plates respectively, and $p_C = n_C k\, T_C$. When $(T_H - T_C)/T_C$ is small, PHILLIPS [2.128] has derived an expression corresponding to linearization of (2.85) and introduction of Knudsen's accommodation coefficients.

DAVIS and ADAIR [2.127] state that (2.85) agrees only approximately with their experimental results. A variety of reasons could explain the difference. However, at $Kn_i \to \infty$, $Kn_L \to \infty$, with the restrictions noted, the state of the gas can be specified exactly; therefore, there is no obvious reason to question the accuracy of (2.85).

In this same limit at sufficiently small host-gas pressures, heating of the particle by thermal radiation can become important and (2.85) would be incorrect. Under these conditions the particle can be said to experience also a photophoretic force. This case has been analyzed approximately by HUGHES [2.130]. The photophoretic force will not be discussed in this survey.

Approximate Descriptions

At present there are no theories or experimental data covering the regimes implied by variation of the full set of dimensionless parameters introduced at the beginning of this section. However, for many practical applications in aerosol physics the regime of primary interest is: $0 \lesssim Kn_i < \infty$, $Kn_L \to 0$, $Ma_{it} \ll 1$, $0 < Br_i \ll 1$.

It will be assumed in the discussion which follows that $Ma_{it} = 0$ because the motion of a particle relative to the temperature gradient changes the measured thermal force as suggested by JACOBSEN and BROCK [2.98] and by PHILLIPS [2.107], whose analysis of this problem will be discussed later. The role of Br_i in the thermal-force problem remains to be investigated. A central point is the separation between mean nonequilibrium and random forces as displayed in (2.45).

Much of the disagreement in experimental and theoretical analysis of the thermal force has been in the slip regime, $0 < Kn_i \lesssim 0.2$. This discussion is directed toward this regime. Sufficient experimental data are not available to permit an assessment of the regimes for $Kn_L > 0$. The various regimes covered by variation of two of the four parameters for the thermal-force problem are displayed in Fig.2.6. The regime displayed as "aerosol data" in this figure will be discussed here.

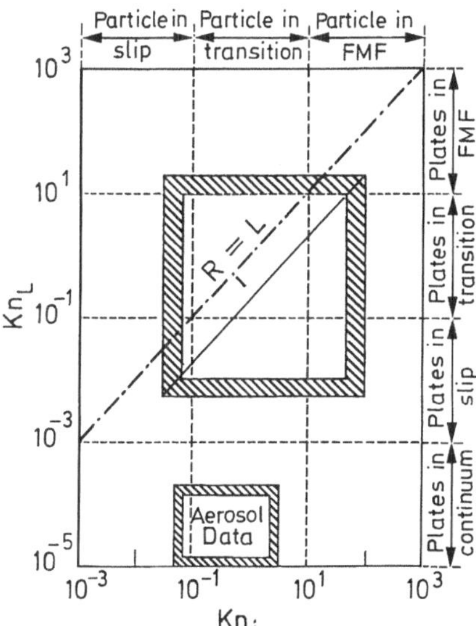

Fig.2.6. Regimes of interest for the thermal force problem in terms of variation of Kn_i and Kn_L. [2.134]

Except for the theories of DERJAGUIN and co-workers [2.121], the theory in this regime has been developed for a stationary particle in a temperature gradient where the thermal force on a particle is exactly balanced by some external force such as an electric field for a charged particle, as in Millikan-cell experiments. The first approximate theory in this case was developed by EPSTEIN [2.131] who used the Navier-Stokes equation with a "thermal creep" boundary condition to obtain:

$$F_{it} = -6\pi\mu_g R_i (k_g/p_0)(2/5)(2 + \Lambda)^{-1} \nabla T_\infty \qquad (2.86)$$

where ∇T_∞ represents the undisturbed temperature gradient at sufficiently large distances from the particle, and p_0 is the average pressure on the particle.

Equation (2.86) displays a marked dependence of F_{it} on the thermal conductivity ratio, Λ, which may range over several orders of magnitude. As experimental data became available for aerosol particle-gas systems with large Λ, it was apparent that (2.86) was seriously in error for the slip-flow regime $0 < Kn_i \lesssim 0.2$, the regime where (2.86) might be expected to be valid.

Several improvements over Epstein's analysis were proposed by BROCK [2.132], JACOBSEN and BROCK [2.98], DWYER [2.133], DERJAGUIN and YALAMOV [2.121], SPRINGER [2.134], PHILLIPS [2.128], and ANNIS and MASON [2.135]. The first four modifications proceeded from various higher-order constitutive relations and boundary conditions and the first three brought about an improvement between theory and Millikan-cell experimental thermal-force data. SPRINGER [2.134] used SUR (2.80) with the Jacobsen-Brock (J-B) equation and Waldmann's equation (2.83) and was able to correlate all existing Millikan thermal-force data for $0 < Kn_i < \infty$. PHILLIPS [2.128] used SUR but with his expression (from Lees' method) for the thermal force instead of the J-B equation and also correlated Millikan-cell data satisfactorily. He enlarged the region of the SUR interpolation to $0 < Kn_i < \infty$, $0 < Kn_L < \infty$. ANNIS and MASON [2.135] used a variant of SUR adapted for their model of an aerosol particle as one component of a multicomponent gas mixture. They claim a region of validity which also includes finite boundaries, $Kn_L > 0$.

From Waldmann's derivation of thermodynamically consistent boundary conditions for higher-order constitutive equations (valid for small Kn_i) [2.103], it appears that the theoretical developments just cited, including those of DERJAGUIN and co-workers, have proceeded from either thermodynamically inconsistent boundary conditions or inaccurate (e.g., Maxwell's) boundary conditions. Therefore, the apparent agreement suggested, for example by SPRINGER [2.134], PHILLIPS [2.128], or ANNIS and MASON [2.135] may be fortuitous, particularly for the slip regime.

For the thermal-force problem VESTNER and WALDMANN [2.103] find the following expression for the thermal force using thermodynamically consistent boundary conditions and constitutive relations which reduce for monatomic gases and Maxwell molecules to Grad's 13 moment equations

$$F_{it} = -6\pi\mu_g R_i (k_g/p_0) \mathscr{A}_t \nabla T_\infty \quad , \qquad (2.87)$$

$$\text{where } \mathscr{A}_t = \left[2(\tilde{\beta} + C_{cr}) + \sum_{q=1}^{4} d_q Kn_i^q\right] / (2 + \Lambda)\left(1 + \sum_{q=1}^{5} b_q Kn_i^q\right) \quad , \qquad (2.88)$$

where the b_q have already appeared in the discussion of the isothermal drag force (2.72) and the d_q are new constants

$$d_1 = \Lambda(C_{cr} + 2\tilde{\beta})C_t - 4(\Lambda - 1)\tilde{\beta}C_m$$

$$d_2 = \tilde{\beta}C_tC_h + 4\Lambda\tilde{\beta}C_tC_m - 8\tilde{\beta}(C_mC_h - \xi C_{cr}^2)$$

$$+ 16\tilde{\beta}^2 C_{cr}\xi + 8\Lambda\tilde{\beta}^3\xi \quad \text{etc.} \quad ,$$

and

$$b_2 = 6C_tC_m[\Lambda/(2 + \Lambda)] + 12C_{cr}\tilde{\beta}\xi(\Lambda + 2)^{-1}$$

$$+ 6(2 + \Lambda)^{-1}(C_mC_h - \xi C_{cr}^2) + C_tC_h/2 + 9\Lambda(2 + \Lambda)^{-1}\tilde{\beta}^2\xi$$

where $\tilde{\beta}$ is a dimensionless coefficient characterizing the Burnett terms in the higher-order constitutive equations. It equals Grad's value of 2/5 for monatomic gases.

In the limit $Kn_i \to 0$, Epstein's equation (2.86) is obtained from (2.87) if it is assumed that $(C_{cr} + \tilde{\beta}) = 1/5$. For $Kn_i \gg 1$, (2.88) becomes

$$\mathscr{A}_t = (4\tilde{\beta}/9)(C_m - \xi C_{cr}^2/C_h)^{-1}Kn_i^{-1} \quad ,$$

which cannot be true if one admits the theoretical possibility that $C_m \gg 1$ see (2.83) in this same limit . However, for near perfect accommodation (2.87) does agree qualitatively with (2.83) in the free-molecular limit.

The difficulty with all the analyses cited lies in the fact that there are at present no a priori methods for determining for a given gas-particle system the phenomenological coefficients C_m, C_t, C_{cr}, C_h, or the various accommodation coefficients which arise for the thermal-force problem. Of course, C_m, C_t, and C_{cr} can be obtained from independent experimental measurements of isothermal drag force, heat transfer, and thermal transpiration, respectively [2.103]. However, C_h is still an adjustable parameter. More importantly from a practical standpoint, the experimental measurements listed may have little relation to the correct values for ultrafine particles. As a consequence, for ultrafine particles the theories contain in reality adjustable parameters which can be made to give good "fits" of available data. In this situation, the virtue of (2.87) is that it rests on correct foundations for the slip regime. Indeed, (2.87) correlates the available thermal-force data very well; in particular (2.87) represents satisfactorily the critical dependence of $\underline{F}_{it}$ on Λ.

Several BGK analyses have been performed for the thermal-creep coefficient [2.99]. The only BGK analysis of the near-free-molecular regime thermal force ($Kn_i \gg 1$, $Kn_L \to 0$) appears to be that of BROCK [2.136]. Experimental Millikan-cell data were found by SCHMITT [2.97] and others to follow the relation

$$F_{it} = F^{*}_{it} \exp(-\tau Kn_i^{-1}) \qquad (2.89)$$

over the range $0 < Kn_i^{-1} \lesssim 6$. BROCK [2.136] modified the BGK analysis to give

$$\tau = 0.09 + 0.12\alpha_t + 0.28\alpha_t(1 - \alpha_h k_g/2k_i) \qquad (2.90)$$

where α_h is the energy accommodation coefficient. Additional experimental thermal force data are necessary to determine the validity of (2.90). Additional analysis also is needed to determine whether or not the next term in the near-free-molecular analysis will include the logarithmic form discussed earlier (2.79).

Experimental Thermal-Force Results for $Kn_i > 0$, $Kn_L \ll 1$

In a discussion of thermal-force theories it is necessary to address the problem of experimental data for the following reason. It has been known for some time that experimental thermal-force data determined by the Millikan-cell method [2.97, 98,137-139] differed from the data obtained by measuring the velocity of particle motion due to the thermal force (thermophoretic velocity) in various flow systems with different configurations [2.121,140,141].

DERJAGUIN and co-workers have offered the following explanation for the differences between their experimental data for the measured thermophoretic velocity and Millikan-cell measurements of thermal force. The values of thermophoretic velocity determined by DERJAGUIN and co-workers are usually from two to four times the corresponding values obtained by the Millikan-cell method. DERJAGUIN and co-workers have claimed that there are uncontrolled convection currents in the Millikan cell and cite the experiments of PARANJPE [2.142] who observed convection currents in the central portion between two parallel plates when the ratio of plate diameter to separation between the plates was less than 5 or 6. There are several reasons to discount this explanation. GIESEKE [2.139] has published experimental thermal-force data in which the ratio of plate diameter to separation, r_a, was varied between 4.9 and 14.9; no effect of variation in r_a was detectable. The data of SCHMITT [2.97] and JACOBSEN and BROCK [2.98] were taken with an r_a of 6.8 and 6.9 respectively. These same data give within 5 percent the theoretical Waldmann limit, (2.83); sufficient data were available at large Kn_i to permit the extrapolation $Kn_i^{-1} \to 0$ with good accuracy. The data of DERJAGUIN and co-workers [2.121] differ from (2.83), being approximately 34 percent too high in the free-molecular limit.

It is still necessary to explain the large differences between the Millikan-cell thermal-force data and the thermophoretic-velocity measurements, principally of DERJAGUIN and co-workers. One possible explanation, which might explain some of the difference, lies in the theory and experiment of PHILLIPS [2.128]. PHILLIPS observed that in Millikan-cell experiments at small Kn_i the thermal-force data for a particle moving under the influence of both gravity and thermal force were some 50 percent greater than the thermal-force data for particles held stationary by means of an electric field. Phillips proposed an approximate slip theory for the effect of velocity on the thermal force. He reports

$$F_{it} = -6\pi\mu_g R_i[(3/4)(2+\Lambda)^{-1}Kn_i\bar{v}_g R_i\{\nabla T_\infty/T + (1/24)R_i^2\underline{V}_i \cdot \nabla T_\infty k_i \underline{V}_i/T(\nu_i\nu_g k_g)\}] \quad (2.91)$$

where ν_i and ν_g are the thermal diffusivities of particle and gas respectively. The first term on the right is Epstein's equation (2.86). The second term indicates the role of the relative velocity of the particle. It can be seen that the measured thermal force (and, therefore, the thermophoretic velocity) is increased so long as the particle velocity has a component either opposed to or in the same direction as the temperature gradient. The explanation is that particle motion changes the temperature gradient in the gas surrounding the particle with a resultant augmentation of the thermal force or thermophoretic velocity. With increasing Kn_i this effect must diminish, since for $Kn_i \to \infty$ the particle does not alter the state of the gas molecules striking the particle. This effect could well explain some of the difference between Millikan-cell thermal-force data for stationary particles and direct measurements of thermophoretic velocity.

It is interesting to note that for $Br_i > 0$, there is a coupling between the diffusive motion and thermophoretic motion of a particle. The relative velocity of a diffusing particle in host gas is

$$\underline{V}_{id} = -D_i \nabla \ln(n_i) \quad (2.92)$$

where D_i is the diffusion coefficient and n_i is the particle concentration. Inserting (2.92) into (2.91) gives for the second term on the right of (2.91)

$$R_i^2 D_i^2[\nabla \ln(n_i) \cdot \nabla T_\infty]k_i \nabla \ln(n_i)/24\nu_i\nu_g k_g T \quad . \quad (2.93)$$

This is only a rough estimate for the slip region. The coupling term is very small and must vanish for $Kn_i \to 0$. A more interesting question concerns the coupling of host-gas fluctuations for small particles in nonequilibrium host gas.

Other Thermal-Force Phenomena

The thermal force discussed heretofore can be attributed to externally imposed momentum transport by the gas molecules. Several theories [2.143,144] have been proposed for an additional effect in which surface-tension gradients are set up in droplets of relatively small viscosity by the temperature gradient. The surface-tension gradient produces internal circulation of the drop. This results in motion of the surrounding gas with a resultant force usually opposite in direction to the thermal force. YALAMOV and SANSARYAN report for the internally driven thermophoretic velocity

$$\underline{u}_{int} = -2(\partial\sigma_i/\partial T_g)(R_i/\mu_g)(1 + \Lambda C_t Kn_i)\nabla T_\infty$$
$$[2 + \Lambda(1 + 2C_t Kn_i)]^{-1}[2\mu_g + \mu_i(1 + 2C_m Kn_i)]^{-1} \quad . \tag{2.94}$$

Since the surface-tension gradient $\partial\sigma_i/\partial T_g < 0$ the motion is opposite to that of the ordinary thermal force, and $\underline{u}_{int} \to 0$ as the viscosity of the particle μ_i becomes much larger than the viscosity of the gas μ_g.

ANNIS and MASON [2.135] have shown that when a particle undergoes both thermophoresis and diffusiophoresis, the two phenomena interact and are each enhanced. These effects have not yet been studied experimentally.

Finally, just as thermophoresis has as a limit thermal diffusion in dilute gas mixtures, so one would expect a thermophoretic effect on particles suspended in dense gases and liquids, whose limit would be thermal diffusion of mixtures in these media. The photophoretic effect may have been observed by BARKAS [2.145] in aqueous solutions of colloids. More recently, McNAB and MEISEN [2.146] have reported experimental evidence of thermophoresis in liquids for 1.011 and 0.79 μm spheres in water and n-hexane. They report that their data for the thermophoretic velocity are described by an empirical equation

$$\underline{u}_{it} = -0.26(2 + \Lambda_\ell)^{-1}(\mu_\ell/\rho_\ell T)\nabla T_\infty \tag{2.95}$$

where the subscript ℓ denotes the liquid phase. McNAB and MEISEN urge caution in using (2.95) because it is based on a single value of Λ_ℓ. They suggest that the observed thermophoresis may be caused by thermal creep, although there is no evidence, except for the similarity of (2.95) to Epstein's equation (2.86). Earlier, the theory of thermophoresis in dense gases and liquids was discussed by BROCK [2.147]. He suggested that thermophoresis of particles in liquids must be caused by both a kinetic part due to molecular impacts and by intermolecular forces acting at a distance between liquid molecules and molecules of the surface of the particle.

Discussion

Much additional analysis and experimental work will be required to place the theory of the thermal force on a rigorous basis. The subject is certainly much more complex than the isothermal drag force discussed earlier. Not only do the regimes $Kn_i > 0$, $Kn_L > 0$ need further investigation, but also the regimes with $Br_i > 0$ and $Ma_{it} > 0$. Very little is known about thermophoresis of nonspherical particles, but it is undoubtedly a complex phenomenon and requires introduction of additional parameters.

investigation of gas-surface interactions is essential. Now, the accommodation coefficients appear in the theory as adjustable parameters for a given gas-particle system.

2.4 Conclusion

As should be evident, this survey has dealt on the whole with idealized models of aerosols and aerosol particles which are different from the complex aerosols of practical interest in such fields as air pollution, cloud physics, aerosol therapeutics, military science, industrial technology, etc. Indeed this complexity presents great difficulties in establishing a science of aerosols.

Even for the simplest problems in aerosol science, such as the approach of a particle to equilibrium in the Brownian-particle approximation, detailed analysis has required treating the particles as rigid spheres or some other axisymmetric shape. Internal states of the particle have been largely ignored. Indeed, for ultrafine particles such states cannot even be inferred from knowledge of the bulk material.

It appears feasible to analyze the homogeneous nucleation process by molecular dynamics for molecules with spherically symmetric potentials, but the next step—extension to nonsymmetric potentials—appears to be almost prohibitive at present. The much more important nucleation processes—heteromolecular nucleation and heterogeneous nucleation, for example—must still be approached by phenomenological theories.

The correct analysis of the evolution of aerosol systems in nonequilibrium host gas has also succeeded only for idealized particle models in the molecular regime. The complexity of evolution in flames, shock waves, etc., can only be understood qualitatively at present.

As pointed out in Sect.2.3, the specification of the gas-particle interaction through the accommodation coefficients presents difficulties. Even for the simple, classical problem of the drag force on a spherical particle in isothermal host gas, current theories are insufficient both as regards the drag calculation from the velocity distribution functions and in the use of boundary conditions. Accommodation coefficients cannot be calculated a priori and solution of the Boltzmann equation throughout the transition regime is still elusive. As Sect.2.3 indicated, the theory of single-particle dynamics in a general nonequilibrium host gas is still incomplete. Additional complication arises when processes such as mass transfer and chemical reaction occur between host gas and particle [2.6,148].

Progress in the theory of ultrafine particles is also impeded in part by difficulties in accurate experimental investigations. For example, as noted in Sect.2.2, experiment in homogeneous nucleation has been unable to resolve differences of 10^{17} between the classical B-D theory and the Lothe-Pound theory.

Currently, inadequate or laborious experimental means are matched by deficient theory. Achievement of a true "aerosol science" will require considerably more innovation in theoretical and experimental methods in the study of the kinetics of ultrafine particles.

Acknowledgments

The author wishes to acknowledge support by the Army Smoke Research Program through the U.S. Army Research Office and by the Aerosol Research Branch, E.S.R.L., U.S. Environmental Protection Agency. The contents do not necessarily reflect the view and policies of the U.S. Army Research Office or the U.S. Environmental Protection Agency.

The author wishes to thank Professor Dr. L. Waldmann and Professor Dr. H. Vestner for their corrections and suggestions regarding the manuscript.

References

2.1 C. Junge: Kolloid Z. Z. Polym. *250*, 638 (1972)
2.2 E. Muetterties: Science *196*, 839 (1977)
2.3 I. Glassman: *Combustion* (Academic Press, New York 1977)
2.4 P. Wegener: In *Nonequilibrium Flows*, Vol.I, ed. by P. Wegener (Dekker, New York 1969)
2.5 N. Fuchs: In *Topics in Current Aerosol Research*, ed. by G.Hidy, J. Brock (Pergamon Press, Oxford 1971)
2.6 G. Hidy, J. Brock: *The Dynamics of Aerocolloidal Systems* (Pergamon Press, Oxford 1970)
2.7 N. Fuchs: *Mechanics of Aerosols* (Pergamon Press, Oxford 1964)
2.8 G. Rudinger: *Nonequilibrium Flows*, Vol.I, ed. by P. Wegener (Dekker, New York 1969)
2.9 N. van Kampen: Adv. Chem. Phys. *34*, 245-308 (1976)
2.10 S. Chandrasekhar: Rev. Mod. Phys. *15*, 61 (1943)
2.11 E. Guth: *Stochastic Processes in Chemical Physics*, ed. by K. Shuler (Interscience, New York 1969)
2.12 S. Nordholm, R. Zwanzig: J. Stat. Phys. *133*, 47 (1975)
2.13 M. Wang, G. Uhlenbeck: Rev. Mod. Phys. *17*, 323 (1945)
2.14 R. Balescu: *Equilibrium and Non-Equilibrium Statistical Mechanics* (Wiley-Interscience, New York 1975)
2.15 H. Mori: Prog. Theor. Phys. *53*, 1617 (1975)
2.16 M. Hoare, C. Kaplinsky: Physica *81*A, 349 (1975)
2.17 K. Andersen, K. Shuler: J. Chem. Phys. *40*, 633 (1964)
2.18 A. Marcus: Technometrics *133* (1964)
2.19 G. Nicolis, I. Prigogine: *Self Organization in Non-Equilibrium Systems* (Wiley, New York 1977)
2.20 E. Mason, S. Chapman: J. Chem. Phys. *36*, 627 (1962)
2.21 S. Chapman, T. Cowling: *The Mathematical Theory of Non-Uniform Gases* (Cambridge University Press, London 1958)
2.22 L. Waldmann: In *Fundamental Problems in Statistical Mechanics*, ed. by E.G.D. Cohen (North-Holland, Amsterdam 1968)
2.23 O. Grechannyi: J. Eng. Phys. *26*, 727 (1974)
2.24 W. Slinn, S. Shen: J. Stat. Phys. *3*, 291 (1971)

2.25 R. Becker, W. Döring: Ann. Phys. (Leipzig) *24*, 719 (1935)
2.26 A. Zettlemoyer (ed.): *Nucleation* (Dekker, New York 1969)
2.27 F. Abraham: *Homogeneous Nucleation Theory* (Academic Press, New York 1974)
A. Albano, D. Bedeaux, P. Mazur: Physica *80*A, 89-97 (1975)
2.28 K. Nishioka, G. Pound: In *Surface and Colloid Science*, Vol.8, ed. by E. Matijevic (Wiley, New York 1976)
2.29 A. Zettlemoyer (ed.): *Nucleation Phenomena* (Elsevier, New York 1977)
2.30 J. Burton: In *Statistical Mechanics*, ed. by B. Berne (Plenum Press, New York 1977) Part A, p.195
2.31 C. Becker, H. Reiss, R. Heist: J. Chem. Phys. *68*, 3585 (1978)
2.32 J. Lothe, G. Pound: J. Chem. Phys. *36*, 2080 (1962)
2.33 V. Bricard, M. Cabane, G. Madelaine: J. Colloid Interface Sci. *58*, 113 (1977)
2.34 A. Sutugin, A. Lushnikov, G. Chernyaeva: Aerosol Sci. *4*, 295 (1973)
2.35 A. Sutugin, N. Fuchs: J. Colloid Interface Sci. *27*, 216 (1968)
2.36 C. Mou: J. Chem. Phys. *68*, 1385 (1978)
2.37 K. Binder, D. Stauffer: Adv. Phys. *25*, 343 (1976)
2.38 K. Kitahara, H. Metiu, J. Ross: J. Chem. Phys. *63*, 3156 (1975)
2.39 B. Berne: In *Aerosol Microphysics, II: Chemical Physics of Microparticles*, ed. by W.H. Marlow, Topics in Current Physics (Springer, Berlin, Heidelberg, New York, in preparation
2.40 W. Zurek, W. Schieve: J. Chem. Phys. *68*, 840 (1978)
2.41 W. Zurek, W. Schieve: Rarefied Gas Dynamics Symposium (to appear)
2.42 W. Schieve, M. Harrison: J. Chem. Phys. *61*, 700 (1974)
2.43 S. Friedlander: J. Colloid Interface Sci. *67*, 387 (1978)
2.44 R. Drake: In *Topics in Current Aerosol Research*, Part 2, ed. by G. Hidy, J. Brock (Pergamon Press, Oxford 1972)
2.45 G. Zebel: In *Aerosol Science*, ed. by C. Davies (Academic Press, New York 1966)
2.46 P. Middleton, J. Brock: J. Colloid Interface Sci. *54*, 249 (1976)
2.47 P. Middleton, J. Brock: APCA J. *27*, 771 (1977)
2.48 T. Peterson, F. Gelbard, J. Seinfeld: J. Colloid Interface Sci. *63*, 426 (1978)
2.49 S. Friedlander: Aerosol. Sci. *1*, 295 (1970)
2.50 F. Gelbard, J. Seinfeld: J. Colloid Interface Sci. *63*, 472 (1978)
2.51 A. Lushnikov: J. Colloid Interface Sci. *54*, 94 (1976)
2.52 P. Wagner, M. Kerker: J. Chem. Phys. *66*, 638 (1977)
2.53 A. Chatterjee, M. Kerker, D. Cooke: J. Colloid Interface Sci. *53*, 71 (1975)
2.54 G. Nicolaon, M. Kerker, D. Cooke, E. Matijevic: J. Colloid Interface Sci. *38*, 460 (1972)
2.55 T. Mercer, M. Tillery: J. Colloid Interface Sci. *37*, 785 (1971)
2.56 S. Devir: J. Colloid Interface Sci. *23*, 80 (1967)
2.57 N. Fuchs, A. Sutugin: J. Colloid Interface Sci. *20*, 492 (1965)
2.58 J. Quon: Int. J. Air Water Pollut. *8*, 355 (1964)
2.59 P. Nolan, E. Kennan: Proc. R. Soc. Irish Acad. *52*A, 171 (1949)
2.60 W. Cawood, R. Whytlaw-Gray: Trans. Faraday Soc. *32*, 1059 (1936)
2.61 C. Jander, A. Winkler: Kolloid Z. *63*, 5 (1933)
2.62 T. Mercer: In *Fundamentals of Aerosol Science*, ed. by D. Shaw (Wiley, New York 1978) Chap.II
2.63 B. Liu, C. Kim: Atmos. Environ. *11*, 1097 (1977)
2.64 S. Loyalka: J. Colloid Interface Sci. *57*, 578 (1976)
2.65 M. Sitarski, J. Seinfeld: J. Colloid Interface Sci. *61*, 261 (1977)
2.66 A. Astakhov: Dokl. Akad. Nauk SSSR *161*, 1114 (1965)
2.67 L. Spielman: J. Colloid Interface Sci. *33*, 562 (1970)
2.68 B. Felderhof: Physica *89*A, 373 (1977)
2.69 J. Deutch, I. Oppenheim: J. Chem. Phys. *54*, 3547 (1971)
2.70 A. Stepanov: In *Hydrodynamics and Thermodynamics of Aerosols*, ed. by V. Voloshehuk, Y. Sedunov (Wiley, New York 1973)
2.71 I. Goodarz-Nia: Chem. Eng. Sci. *33*, 533 (1978)
2.72 I. Goodarz-Nia: J. Colloid Interface Sci. *52*, 29 (1975)
2.73 I. Goodarz-Nia, D. Sutherland: Chem. Eng. Sci. *30*, 407 (1975)
2.74 D. Sutherland, I. Goodarz-Nia: Chem. Eng. Sci. *26*, 2071 (1971)

2.75 D. Sutherland: J. Colloid Interface Sci. *25*, 373 (1967)
2.76 A. Medalia, F. Heckman: Carbon *7*, 567 (1969)
2.77 J. Beeckmans: Ann. Occup. Hyg. *7*, 299 (1964)
2.78 M. Vold: J. Colloid Interface Sci. *18*, 684 (1963)
2.79 H.R. Kruyt: *Colloid Science*, Vol.I (Elsevier, Amsterdam 1952)
2.80 W. Stöber, H. Flaschbart, D. Hochrainer: Staub *30*, 277 (1970)
2.81 W. Stöber: In *Assessment of Airborne Particles*, ed. by T. Mercer et al. (Thomas, Springfield 1972)
2.82 J. Kops, G. Dibbets, L. Hermans, J. Van de Vate: Aerosol Sci. *6*, 329 (1975)
2.83 L. Waldmann: In *Aerosol Science*, ed. by C. Davies (Academic Press, New York 1966)
2.84 C. Cercignani: *Theory and Application of the Boltzmann Equation* (Elsevier, New York 1975)
2.85 J. Friedel: J. Phys. *38*, Colloq. C-2 (1977)
2.86 F. Goodman, H. Wachman: *Dynamics of Gas-Surface Scattering* (Academic Press, New York 1976)
2.87 J. Maxwell: Philos. Trans. R. Soc. (London) *170*, 231 (1879)
2.88 I. Kuscer: In *Theoriques Cinetiques Classiques et Relativistes* (C.N.R.S., Paris 1975)
2.89 L. Waldmann: Z. Naturforsch. *22*a, 1269 (1967)
2.90 L. Waldmann, H. Vestner: Physica *80*A, 523 (1975)
2.91 I. Kuscer, T. Klinc: Phys. Fluids *15*, 1018 (1972)
2.92 L. Waldmann: Z. Naturforsch. *32*a, 521 (1977)
2.93 A. Basset: Philas. Trans. R. Soc. London *179*, 43 (1888)
2.94 P. Epstein: Phys. Rev. *23*, 710 (1924)
2.95 M. Kogan: In *Annual Reviews of Fluid Mechanics*, Vol.5, ed. by M. Van Dyke, W. Vincenti, J. Wehausen (Annual Reviews, Palo Alto 1973) p.383
2.96 R. Millikan: Phys. Rev. *221* (1924)
2.97 K. Schmitt: Z. Naturforsch. *14*a, 870 (1959)
2.98 S. Jacobsen, J. Brock: J. Colloid Sci. *20*, 544 (1965)
2.99 S. Loyalka, N. Petrellis, T. Storvick: Phys. Fluids *18*, 1094 (1975)
2.100 H. Mott-Smith: Phys. Rev. *82*, 885 (1951)
2.101 H. Grad: Commun. Pure Appl. Math. *4*, 331 (1949)
2.102 L. Lees: GALCIT Res. Proj., Memo 51 (1959)
2.103 H. Vestner, L. Waldmann: Physica *86*A, 303 (1977)
2.104 L. Lees: J. Appl. Math. Phys. *28*, 835 (1977)
2.105 R. Goldberg: Ph.D. Thesis, New York University (1964)
2.106 C. Liu, T. Sugimura: In *Rarefield Gas Dynamics*, ed. by L. Triling (Academic Press, New York 1969)
2.107 W. Phillips: Phys. Fluids *18*, 1089 (1975)
2.108 Y. Sone, K. Aoki: Phys. Fluids *20*, 571 (1977)
2.109 Y. Pao, D. Willis: Phys. Fluids *12*, 435 (1969)
2.110 J. Dorfman, H. Van Beijeren, C. McClure: Arch. Mech. *28*, 333-352 (1973)
2.111 Y. Kan: Ph.D. Thesis, Dept. of Physics and Astronomy, University of Maryland (1975)
2.112 D. Willis: University of California, Aero. Sci. Report No. A5-65, 16 (1965)
2.113 F. Sherman: In *Rarefield Gas Dynamics*, Vol.II, ed. by J. Laurmann (Academic Press, New York 1963) p.288
2.114 B. McCoy, C. Cha: Chem. Eng. Sci. *29*, 381 (1974)
2.115 B. Annis, A. Malinauskas, E. Mason: Aerosol. Sci. *3*, 55 (1972)
2.116 W. Stöber: Aerosol Sci. *2*, 453 (1971)
2.117 I. Gallily, A. Cohen: J. Colloid Interface Sci. *56*, 443 (1976)
2.118 D. Hochrainer, G. Hänel: Aerosol Sci. *6*, 97 (1975)
2.119 B. Dahneke: Aerosol Sci. *4*, 139 (1973)
2.120 J. Tyndall: Proc. R. Inst. *6*, 3 (1870)
2.121 B. Derjaguin, Y. Yalamov: In *International Reviews in Aerosol Physics and Chemistry*, Vol.3, ed. by G. Hidy, J. Brock (Pergamon Press, New York 1972)
2.122 H. Green, W. Lane: *Particulate Clouds*, 2nd ed. (Van Nostrand, New York 1964)
2.123 L. Waldmann: In *8th Rarefied Gas Dynamics Symposium*, ed. by K. Keremcheti (Academic Press, New York 1974) pp.431-449
2.124 L. Waldmann: Z. Naturforsch. *14*a, 589 (1959)

2.125 S. Bakanov, B. Derjaguin: Discuss. Faraday Soc. *30*, 130 (1960)
2.126 L. Monchick, Y. Yun, E. Mason: J. Chem. Phys. *39*, 654 (1963)
2.127 L. Davis, T. Adair: J. Chem. Phys. *62*, 2278 (1975)
2.128 W. Phillips: Phys. Fluids *15*, 999 (1972)
2.129 N. Tong: J. Colloid Interface Sci. *51*, 143 (1975)
2.130 W. Hughes: J. Colloid Sci. *15*, 307 (1960)
2.131 P. Epstein: Z. Phys. *54*, 537 (1929)
2.132 J. Brock: J. Colloid Interface Sci. *17*, 768 (1962)
2.133 H. Dwyer: Phys. Fluids *10*, 976 (1967)
2.134 G. Springer: J. Colloid Interface Sci. *34*, 215 (1970)
2.135 B. Annis, E. Mason: Aerosol Sci. *6*, 105 (1975)
2.136 J. Brock: J. Colloid Interface Sci. *25*, 465 (1967)
2.137 P. Rosenblatt, V. La Mer: Phys. Rev. *70*, 387 (1946)
2.138 O. Schadt, K. Cadle: J. Colloid Sci. *65*, 1689 (1961)
2.139 J. Gieseke: J. Air Pollut. Control Assoc. *18*, 682 (1968)
2.140 E. Keng, C. Orr: J. Colloid Interface Sci. *22*, 107 (1966)
2.141 P. Goldsmith, F. May: In *Aerosol Science*, ed. by C. Davies (Academic Press, New York 1966) Chap.VII
2.142 M. Paranjpe: Proc. Indian Acad. Sci. *40*, 423 (1936)
2.143 G. Gardner: Aerosol Sci. *6*, 173 (1975)
2.144 Y. Yalamov, A. Sanasaryan: Sov. Phys. Tech. Phys. *20*, 1351 (1976)
2.145 W. Barkas: Philos. Mag. *9*, 7 (1930)
2.146 G. McNab, A. Meisen: J. Colloid Interface Sci. *44*, 339 (1973)
2.147 J. Brock: Proc. Symp. on the Less Common Means of Separation, April 1963 (Hodgson, London 1963)
2.148 M. Kogan: *Rarefield Gas Dynamics* (Plenum Press, New York 1969)

3. Classical and Statistical Theories of Gas-Surface Energy Transfer

J. D. Doll

With 11 Figures

Energy transfer between a gas and some other system is a process of interest to workers in a number of areas. Depending on whether the second system is another gas molecule, a small aggregate of molecules, or a solid surface; gas-phase kinetics, aerosol microphysics, or solid-state studies are involved. Energy-transfer theories are best developed for the gas-phase case where currently available methods provide reasonably reliable treatment for both the interactions and the collision dynamics.

The present work attempts to develop an energy-transfer theory suitable for the treatment of condensed-phase systems. The present discussion will concentrate explicitly on gas-solid systems since the development is most straightforward there. Both the approach and the numerical methods have broader applicability, however. We do not intend the following to be an exhaustive review of gas-solid energy-transfer methods since a number of excellent reviews already exist [3.1-3].

The approach described here is an outgrowth of earlier work by ZWANZIG [3.4] and GOODMAN [3.5]. The essential feature of the approach is the utilization of modern nonequilibrium statistical-mechanical methods to simplify the dynamics. Such a step is important since the high level of detail required in these calculations plus the large number of degrees of freedom involved would otherwise render numerical simulations effectively impossible. For the present problem it proves possible to include the effects of those particles not immediately involved in the collision without having to explicitly follow their dynamics. These "background" particles become a generalized solvent for the particles of primary interest. Development of efficient computational procedures based on the present approach is possible.

The present work does not come to grips with the ab initio determination of the interaction potentials involved. The logical development of dynamical and electronic structure methods are, however, to a large extent independent. Furthermore, dynamical studies can be utilized to establish empirically the interaction potentials, or, at least, to establish those features which are most important to the process of interest. Consequently, it is useful to continue the development of dynamical methods even in the absence of (dynamically) useful interaction potentials.

Section 3.1 introduces a number of concepts useful for the discussion of energy transfer by considering the simple driven-oscillator model. Section 3.2 briefly sketches the direct, molecular dynamics approach to gas-surface collisions. The present approach and some sample results are discussed in Sects.3.3,4, respectively.

3.1 A Simple Example: Driven-Oscillator Model

In order to establish certain general features of gas-surface energy transfer it is helpful to consider a simple driven-oscillator model of the collision [Ref.3.1, Chap.10] and [3.6-8]. Although ultimately too simplistic (for example, the model is restricted to collinear collisions), this model provides a qualitatively useful picture of energy transfer and thermal accommodation. The model is shown schematically in Fig.3.1. A gas atom of mass m_g and energy E_g is incident upon a surface oscillator of mass m_s and frequency ω_0. The effects of the remainder of the lattice are represented by a frictional force acting on the surface atom. At this stage the use of frictional terms to simulate the background lattice is simply an assumption. Section 3.3 will consider this point in detail. The classical equations of motion for the two particles are

$$m_s\ddot{x}_s = -m_s\omega_0^2 x_s - \beta_0\dot{x}_s + \frac{\partial V}{\partial X_G}(x_g - x_s) \tag{3.1}$$

$$m_g\ddot{x}_g = -\frac{\partial V}{\partial x_g}(x_g - x_s) \quad , \tag{3.2}$$

where it is assumed that the gas-surface interaction potential is a function only of $x_g - x_s$. Closed form solutions of (3.1,2) for arbitrary interaction potentials are not available. However, a useful approximation procedure involves setting $x_s = 0$ in both terms involving $V(x_g - x_s)$. This approximation decouples the determination of $x_g(t)$ from that of $x_s(t)$. The gas motion is fixed by the elastic trajectory determined by $V(x_g)$ (i.e., with the surface atom frozen at its equilibrium position). The surface-atom motion is in turn fixed by (3.1) where it is now assumed that the gas particle supplies a *fixed* time-dependent external driving force. Physically this approximation asserts that the time-dependent force generated by the gas atom moving on the "frozen-lattice" trajectory is typical of the actual forces that would be generated if (3.1,2) were solved exactly. As is shown below, the error made by neglecting the reaction of the surface motion on the gas particle's trajectory can be estimated in a rather simple way.

One potential for which the above prescription can be carried out analytically is the Morse interaction

$$V(x) = D[\exp(-2\alpha x) - 2\exp(-\alpha x)] \quad . \tag{3.3}$$

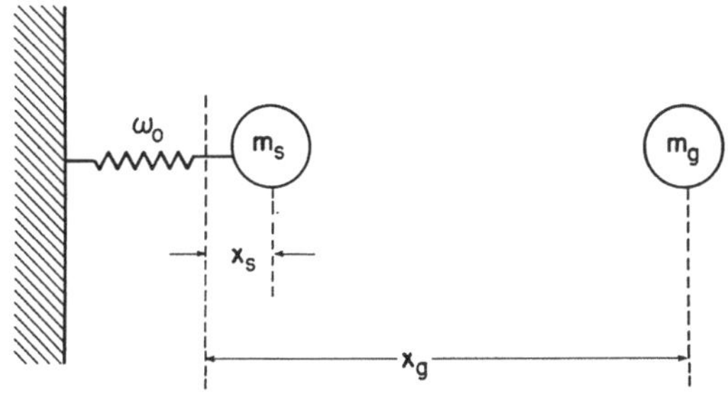

Fig.3.1. The collinear-atom-oscillator model. The gas and oscillator masses are m_g and m_s, respectively. The distances from the oscillator equilibrium position of the gas and surface atom are x_g and x_s. The oscillator frequency is ω_0

The frozen-lattice trajectory of the gas atom is thus fixed by the orbit equation

$$t = \int_{x_0}^{x} dx'\{2[E_g - V(x')]/m_g\}^{-\frac{1}{2}} \quad . \tag{3.4}$$

We assume that at t = 0 the gas is at its turning point, x_0, where $V(x_0) = E_g$. Defining some useful constants

$$d \equiv D/E_g \tag{3.5}$$

$$\omega_c = \alpha\sqrt{2E_g/m_g} \quad , \tag{3.6}$$

and making the substitution

$$u(t) = \exp[-\alpha x(t)] \quad , \tag{3.7}$$

(3.4) becomes

$$\omega_c t = -\int_{u_0}^{u} dv\, v^{-1}[1 - d(v^2 - 2v)]^{-\frac{1}{2}} \quad . \tag{3.8}$$

When integrated this becomes

$$u(t) = [\sqrt{d^2 + d}\ \cosh(\omega_c t) - d]^{-1} \quad , \tag{3.9}$$

or

$$x(t) = \alpha^{-1}\ \ln[\sqrt{d^2 + d}\ \cosh(\omega_c t) - d] \quad . \tag{3.10}$$

Thus the time-dependent force exerted by the gas particle on the surface atom during its frozen-lattice trajectory is

$$f(t) = \frac{\partial V[x_g(t)]}{\partial x_g(t)} \quad ,$$

or

$$f(t) = 2\alpha D[(1+d) - \sqrt{d^2+d}\cosh(\omega_c t)]/[\sqrt{d^2+d}\cosh(\omega_c t) - d]^2 \quad . \qquad (3.11)$$

Where only the repulsive term of the potential in (3.3) present, the result at this stage of the analysis would be the $d = 0$ limit of (3.11),

$$f(t) = -2\alpha E\,\mathrm{sech}^2(\omega_c t) \quad . \qquad (3.12)$$

Since the response of the oscillator will involve the frequency components of the above force, we need the Fourier transform of $f(t)$, given by

$$F(\omega) = \int_{-\infty}^{\infty} \exp(-i\omega t) f(t) dt \quad . \qquad (3.13)$$

Since $f(t)$ is an even function of t, (3.13) reduces to

$$F(\omega) = 2\int_{0}^{\infty} f(t)\cos(\omega t) dt \quad . \qquad (3.14)$$

Although tedious this can be evaluated for (3.11) ultimately yielding

$$F(\omega) = [-\pi m_g \omega/\alpha \sinh(0.5\pi\omega_c^{-1}\omega)]$$
$$\times \{\cosh[\omega_c^{-1}\omega \cos^{-1}(-(-1+d^{-1})^{-\frac{1}{2}})]\}/\cosh(0.5\pi\omega_c^{-1}\omega) \quad . \qquad (3.15)$$

For (3.12) the result is given by the leading (bracketed) term in (3.15), indicating that with respect to $F(\omega)$ the effect of the attractive well is a multiplicative one. From (3.15) we see that the important parameters in $F(\omega)$ are the ratio of the frequency ω to the collision frequency $[\omega_c = (\alpha^{-1}/\sqrt{2E_g/m_g})^{-1}]$ and the ratio of the well depth to the incident energy. Shown in Fig.3.2 are sample plots of $|F(\omega)|$ for various system parameters. We note that for large d values there is a pronounced peak in $|F(\omega)|$ near the frequency corresponding to low-amplitude vibration near the bottom of the Morse well, $\omega = \alpha\sqrt{2D/mg}$. As d is decreased the influence of the well diminishes while the effect of translational motion ($\omega \sim 0$) is enhanced.

As is discussed by MARTIN [3.9] in detail, the response of an oscillator to an external, time-dependent force is straightforward. If the oscillator equation of motion is

$$m_s\ddot{x}_s = -m_s\omega_0^2 x_s - \beta_0 x_s + f(t) \quad , \qquad (3.16)$$

then the Fourier transform of $x_s(t)$

$$X_s(\omega) = \int_{-\infty}^{\infty} x_s(t)\exp(-i\omega t) dt \qquad (3.17)$$

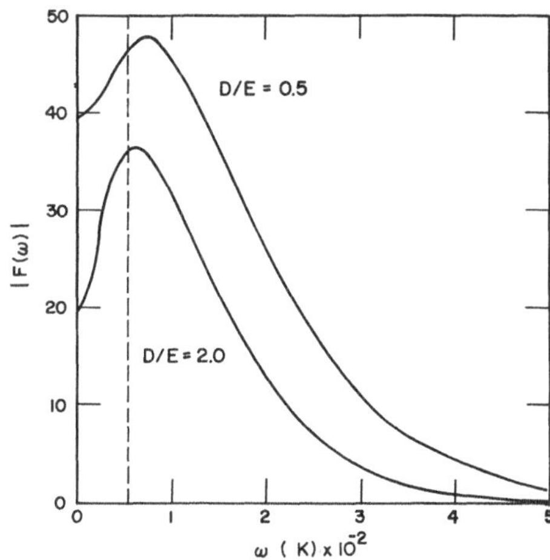

Fig.3.2. Shown for two D/E ratios are the frequency components of the force exerted by the gas on the surface oscillator when the gas moves on the frozen lattice trajectory (3.15). Parameters are D = 418 K, α = 1.69 Å^{-1}, m_g = 39.95

is given by

$$X_s(\omega) = \chi(\omega)F(\omega) \tag{3.18}$$

where

$$\chi(\omega) = [m_s(\omega_0^2 - \omega^2 + i\omega\beta_0)]^{-1} \quad . \tag{3.19}$$

Ultimately this means that the energy transfer to the oscillator, taken to be initially at rest, is given by

$$\Delta E = \int_0^\infty d\omega[-\omega\chi''(\omega)/\pi]|F(\omega)|^2 \tag{3.20}$$

where $\chi''(\omega)$ is the imaginary part of $\chi(\omega)$. For the simple friction model (3.19)

$$\chi''(\omega) = -\omega\beta_0/\{m_s[(\omega_0^2 - \omega^2)^2 + (\omega\beta_0)^2]\} \quad . \tag{3.21}$$

An important limiting case of (3.20) is for β_0 = 0. In the limit of vanishing friction $\omega\chi''(\omega)$ becomes a delta function centered at ω_0. Thus E becomes

$$\Delta E = |F(\omega_0)|^2/2m_s \quad . \tag{3.22}$$

Using (3.15) and various hyperbolic identies this becomes

$$\Delta E = 4\mu E_g\{\pi\xi \cosh[\xi\cos^{-1}(-(-1 + d^{-1})^{-\frac{1}{2}})]/\sinh(\pi\xi)\}^2 \quad . \tag{3.23}$$

where

$$\mu = m_g/m_s \tag{3.24}$$

$$\xi = \omega_0/\omega_c \quad . \tag{3.25}$$

Dividing both sides of (3.23) by E_g we obtain the driven-oscillator result for the thermal-accommodation coefficient (at a fixed energy)

$$\alpha_E = 4\mu\{\pi\xi \cosh[\xi\cos^{-1}(-(1 + d^{-1})^{-\frac{1}{2}})]/\sinh(\pi\xi)\}^2 \quad . \tag{3.26}$$

It is common when using (3.26) or related expressions to assume that α_E is independent of the surface temperature.

The driven-oscillator model reveals several interesting points concerning the behavior of thermal-accommodation coefficients. We first note that (3.26) predicts that the accommodation coefficient is a universal function of three dimensionless parameters, the mass ratio, μ, the reduced well depth, d, and the adiabaticity parameter, ξ. This latter quantity is the ratio of the duration of the collision to the oscillator period. Abrupt, impulsive collisions correspond to ξ values near zero, while long lasting, languid collisions correspond to large ξ values. For small ξ values we obtain from (3.26)

$$\lim_{\xi \to 0} \alpha_E = 4\mu \quad , \tag{3.27}$$

indicating that under these conditions the only relevant quantity is the mass ratio. Equation (3.27) has been derived by considering the impulsive collision as the limiting case of a driven-oscillator model. However, the impulsive-collision situation can be treated directly. The result is

$$\alpha_E(\xi = 0) = 4\mu/(1 + \mu)^2 \quad . \tag{3.28}$$

The difference between (3.27,28) is that in obtaining (3.28) we have allowed the trajectory of the incident particle to "readjust" its motion during the collision event. This suggests that we can to some extent correct the results derived from (3.20) and (3.15) to remove the effects of the frozen-lattice assumption by a simple renormalization. That is, we could modify (3.26) [and (3.23)] by requiring that the results match the known impulsive limits. Equation (3.26) thus becomes

$$\alpha_E = [4\mu/(1 + \mu)^2]\{\pi\xi \cosh[\xi \cos^{-1}(-(1 + d^{-1})^{-\frac{1}{2}})]/\sinh(\pi\xi)\}^2 \quad . \tag{3.29}$$

As is commonly the case in energy-transfer results [3.10] (3.30) reveals a "resonance-function" behavior. That is, $\alpha_E(\xi,d)$ is of the form

$$\alpha_E(\xi,d) = \alpha_E(\xi = 0)R(\xi,d) \quad . \tag{3.30}$$

Thus, as a function of ξ, α_E is the impulsive value scaled by some functions of ξ (and in this case d). The softness of the interaction potential considerably diminishes the value of α_E compared with the hard-cube result (3.28), at least in the range $E_g > D$.

The presence of an attractive well in the interaction potential produces an interesting effect in (3.29). As E_g is reduced from very high values (impulsive collisions) the initial effect is to make the collisions less abrupt and hence to reduce α_E. As E_g is further reduced to values comparable to D, however, the well exerts an increasingly large effect on the collision. In particular, it prevents the collision from being arbitrarily languid. Thus for small E_g (large d and ξ) we see from (3.29) that α_E behaves as $1/E_g$. This implies that as a function of E_g, α_E has a minimum in the vicinity of $E_g = D$. This minimum is connected to the presence of an attractive well, but is not necessarily associated with the onset of trapping. Figure 3.3 shows for a prototype system $\alpha_E(E_g)$ values predicted by (3.29). We do see a minimum in the vicinity of D.

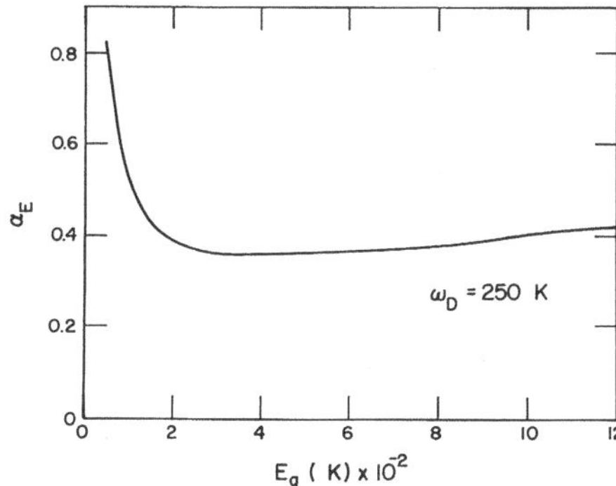

Fig.3.3. The driven-oscillator values for α_E (3.29). The system parameters are those of Fig.3.2 with ω_0 = 194 K and m_s = 108

The $1/E_g$ increase of α_E for E_g values near zero predicted by (3.30) can not constitute a physical singularity. Trapping will cut off α_E at a value of unity. A simple estimate of the critical trapping energy, E_g^*, the energy which locates the onset of trapping, can be obtained from (3.30) by setting

$$\alpha_E(E = E_g^*) = 1 \quad . \tag{3.31}$$

The above discussion easily generalizes to the case where the frictional term in the oscillator equation of motion becomes time dependent. If (3.16) is of the form

$$m_s\ddot{x}_s = -m_s\omega_0^2x_s - \int_{-\infty}^{t} \beta(t-t')\ddot{x}_s(t')dt' + f(t) \qquad (3.32)$$

(3.18,20) remain valid if $\chi(\omega)$ is taken as

$$\chi(\omega) = \{m_s[\omega_0^2 - \omega^2 + i\omega\beta(\omega)]\}^{-1} \quad . \qquad (3.33)$$

In deriving (3.33) we use the fact that $\beta(t)$ vanishes for $t < 0$. Thus $\beta(\omega)$ is given by

$$\beta(\omega) = \int_0^{\infty} \beta(t)\exp(-i\omega t)dt \quad . \qquad (3.34)$$

Separating $\beta(\omega)$ into its real and imaginary components, $\beta'(\omega)$ and $\beta''(\omega)$, $\chi''(\omega)$ becomes

$$\chi''(\omega) = -\omega\beta'(\omega)/\{m_s[(\omega_0^2 - \omega^2 - \omega\beta''(\omega))^2 + (\omega\beta'(\omega))^2] \quad . \qquad (3.35)$$

For later discussions it will be useful to note that a connection exists between $\chi''(\omega)$ and the distribution of frequencies in our oscillator system. Equation (3.22) specifies the energy transfer to a single oscillator of frequency ω_0. Were the external force acting uniformly on a system of oscillators whose frequency distribution was specified by $g(\omega)$[1], the energy transfer would be given by

$$\Delta E = \int_0^{\infty} 0.5g(\omega)|F(\omega)|^2/m_s d\omega \quad . \qquad (3.36)$$

Comparison with (3.20) shows that the mode density and $\chi''(\omega)$ are essentially proportional. By assuming simple functional forms for $g(\omega)$ we can use (3.36) to generate analogs of the single-oscillator results for energy transfer, accommodation coefficients, and critical trapping energies. For example, if we use a Debye distribution

$$g(\omega) = \begin{cases} 3\omega^2/\omega_D^3 & \omega \leq \omega_D \\ 0 & \omega > \omega_D \end{cases} \qquad (3.37)$$

we find the (renormalized) accommodation coefficient is given in terms of the parameters discussed above by

[1]The normalization is $\int g(\omega)d\omega = 1$.

$$\alpha_E = 4\mu(1+\mu)^{-2} I(\omega_D/\omega_c, d) \tag{3.38}$$

where

$$I(\omega_D/\omega_c, d) = 3(\omega_c/\omega_D)^3 \int_0^{\omega_D/\omega_c} x^2\{\pi x \cosh[x \cos^{-1}(-(1+d^{-1})^{-\frac{1}{2}})]/\sinh(\pi x)\}^2 dx \quad . \tag{3.39}$$

The Debye model, first proposed by ADELMAN and DOLL [3.11] in a different context, is useful since it makes connection with the lattice Debye temperature. Equation (3.39) again suggests a "corresponding states" form for α_E involving the ratios m_g/m_s, ω_D/ω_c, and D/E_g. Similar results for the energy transfer and critical trapping energy can be derived.

The present models, although useful, are not of sufficient generality. The assumptions utilized to simplify the collision dynamics are severe and must ultimately be discarded. This means that the collision dynamics must be simulated numerically. The outline of such an approach is discussed below.

3.2 Classical Mechanical Treatment of Gas-Solid Collisions

While no one presently doubts that quantum mechanics forms the general theoretical framework for the discussion of molecular processes, the question of what is the most convenient computational algorithm for certain classes of problems is open. For problems involving heavy-particle dynamics and in situations where appreciable averaging is involved, classical or semiclassical methods [3.12,13] often prove acceptable. For these problems it is often the case that classical mechanics provides a poor picture of the elementary aspects of the process (e.g., specific transition probabilities) while at the same time it gives a rather good estimate of more-averaged quantities (e.g., energy transfer).

A simple example of this ability of classical mechanics to reproduce averaged quantities is illustrated by the model of energy transfer in a collinear hard-sphere-harmonic-oscillator collision discussed by EASTES et al. [3.14]. This model is useful in that the simple dynamics permits closed-form classical (and semi-classical) results. Shown in Table 3.1 are the classically and quantum mechanically computed energy transfers to an oscillator initially in its ground state as a function of the incident gas energy. The classical results are quite reasonable, while the classical values of specific transition probabilities in this energy range are rather poor. In particular, for the parameters used in Table 3.1, all transitions out of the ground state are classically forbidden [3.13] below an

Table 3.1. Shown are the average energy transfers for the collinear hard-sphere-atom/harmonic-oscillator model computed classically (CM) and quantum mechanically (QM). The energies are in terms of multiples of the oscillator spacing. The atom/oscillator mass ratio was 0.02 in this problem

Incident energy	ΔE_{CM}	ΔE_{QM}
1.5	0.077	0.067
2.5	0.154	0.149
3.5	0.231	0.228
4.5	0.308	0.305
5.5	0.385	0.382

incident energy of 3.86 times the oscillator frequency [Ref.3.14, Eq.65]. This tendency of classical mechanics to reproduce quantities involving appreciable averaging is a general one and suggests that for heavy-particle gas-surface dynamics the method will be adequate for many purposes.

We note in passing that by adopting a semiclassical viewpoint [3.13] it is often possible to extend an essentially classical framework to the point that it provides a quantitative treatment of the more elementary features of the collision dynamics.

Before explicitly considering classical mechanical applications to gas-surface dynamics, it is useful to note the following general structure of the results. If the probability distribution function for some quantity, x, is known, P(x), and some function of x is formed, f(x), then the distribution function for f, g(f), is given by

$$g(f) = \int dx P(x) \delta[f(x) - f] \quad . \tag{3.40}$$

In the delta function f (without) with an argument denotes (a specified numerical value) the function's value at the point x. Since the delta function vanishes except at the roots of the equation

$$f(x) = f \quad , \tag{3.41}$$

the integral in (3.40) can be performed by expanding f(x) about these roots. That is, in the vicinity of the ith root of (3.41), x_i, we write

$$f(x) \sim f + \left(\frac{df}{dx}\right)_i (x - x_i) \tag{3.42}$$

where $(df/dx)_i$ implies that the derivative is evaluated at $x = x_i$. Thus the integral in (3.40) yields the familiar result [3.15]

$$\begin{aligned} g(f) &= \sum_i \int dx P(x) \delta\left[\left(\frac{df}{dx}\right)_i (x - x_i)\right] \\ &= \sum_i P(x_i)/|df/dx|_i \quad . \end{aligned} \tag{3.43}$$

The generalization of (3.43) to several functions, f_1, f_2, ..., f_n, of several variables, x_1, x_2, ..., x_n, is

$$g(f_1,f_2,\ldots,f_n) = \sum_i P(x_{1i},x_{2i},\ldots,x_{ni}) \Bigg/ \left| \frac{\partial(f_1,f_2,\ldots,f_n}{\partial(x_1,x_2,\ldots,x_n} \right|_i \tag{3.44}$$

where the Jacobian is evaluated of the roots of the set of equations

$$\begin{aligned} f_1(x_1,x_2,\ldots,x_n) &= f_1 \\ f_2(x_1,x_2,\ldots,x_n) &= f_2 \\ &\vdots \\ f_n(x_1,x_2,\ldots,x_n) &= f_n \quad . \end{aligned} \tag{3.45}$$

This Jacobian structure carries over to classical collision theory, the link between initial and final conditions being a dynamical one. As a concrete example consider the case of a monoenergetic beam of particles incident upon a rigid surface with a periodic structure. For simplicity we restrict the discussion to in-plane scattering. The relevant quantities are defined in Fig.3.4. Since we assume the incident beam uniformly covers the unit cell (of length a), the probability distribution of x_0 (cf Fig.3.4) is

$$P(x_0) = 1/a \quad .[2] \tag{3.46}$$

The final angle, θ_f, as a function of x_0 is schematically depicted in Fig.3.5a. From (3.43) the distribution function for final angles is

$$P(\theta_f) = a^{-1} \sum_i [|d\theta_f/dx_0|_i]^{-1} \tag{3.47}$$

where the sum is over roots of the equation

$$\theta_f(x_0) = \theta_f \quad . \tag{3.48}$$

Thus the probability distribution is connected to the reciprocal of the slope of the $\theta_f(x_0)$ curve at those values of x_0 which lead to the desired final angle. Shown schematically in Fig.3.5b is the θ_f distribution produced by (3.47). Extrema

---

[2]Normalization is such that $\int_0^a P(x_0)dx_0 = 1$.

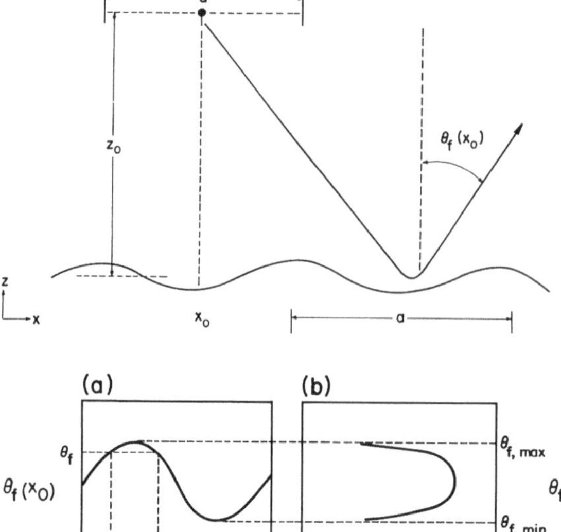

Fig.3.4. Geometry for in-plane scattering from a periodic surface. Initial gas position is (x_0,z_0), where x_0 is uniformly distributed over unit cell and z_0 is large enough that the gas-surface potential is negligible

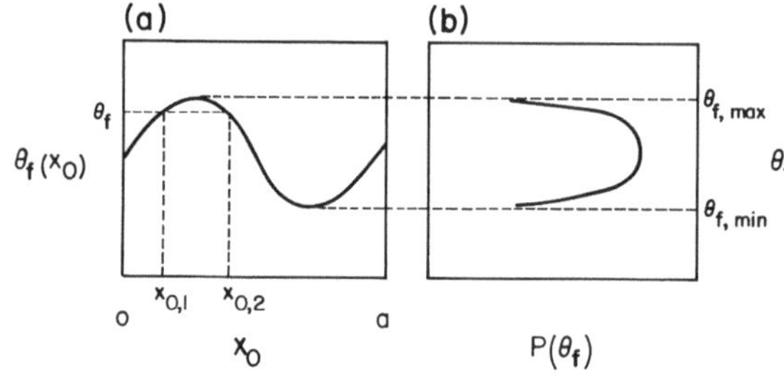

Fig.3.5. (a) Typical $\theta_f(x_0)$ curve for system in Fig.3.4, and (b) the associated probability distribution for θ_f (3.47)

in the $\theta_f(x_0)$ curve induce "rainbow singularities" in the θ_f distribution since the Jacobian in (3.47) vanishes at these points. Values of θ_f beyond these rainbow limits are "classically forbidden," meaning that there are no (real valued) solutions of (3.48) for these regions.[3] Similar considerations apply to problems involving larger numbers of degrees of freedom.

Modern use of classical methods to model molecular-collision processes had its origin in gas-phase collision studies (see [3.12] for a review). The methods have become a mainstay in this field, both because of their relative simplicity and because of their (often surprising) quantitative nature. Recent applications to the prediction of IR spectra [3.16] and to laser induced photodecomposition [3.17,18] give some idea of the present scope of the methods.

Applications of classical methods to gas-surface dynamics have been considered by a number of authors. Problems of energy transfer and scattering distributions have been considered by GOODMAN [3.1,5], ZWANZIG [3.4], McCLURE [3.19-22], OMAN [3.23-26], and RAFF et al. [3.27-30]. More recently GELB and CARDILLO [3.31,32] and McCREERY and WOLKEN [3.33-35] have considered prototype reactive systems involving dissociation and recombination on solid surfaces. The basic approach in all of these studies involves postulating models for both the gas-surface inter-

[3]See [3.13] for a discussion of the implication of the existence of complex roots of (3.48).

action potential and for the lattice characteristics. Use is made of any experimental information available, but it is typically the case that the models involve a variety of semiempirical parameters. The dynamics of the chosen process are then followed by numerically integrating the classical equations of motion for the gas-surface system. Sample trajectories, generated for suitable chosen gas and lattice initial conditions, allow the construction of desired quantities. A best-fit procedure matching available experimental data and theoretical predictions allows the extraction of empirical gas-surface interaction-potential and lattice parameters. These studies are encouraging to the extent that "reasonable" choices for the models produce reasonable agreement with available experimental results, provided a sufficiently large trajectory sample is employed [3.21,22].

Although the molecular-dynamics techniques sketched above will continue to be useful in a wide variety of applications, they do have some important shortcomings. Reasonably converged results often require rather large trajectory samples (e.g., $\sim 10^4$ for the studies reported by McCLURE. Although the dynamics in these problems is straightforward to simulate, the necessity for such large samples means their studies have rather large computing requirements. This is particularly true since calculations with many different parameter sets are often desired. A related defect is that because of the computing requirements, models for the lattice are often oversimplified. This poses the risk that any derived quantities may be biased by the artificial aspects of the model. It is often the case that a nonphysical model can produce reasonable agreement with a limited set of experimental data if the model's parameters are allowed sufficient flexibility. At a more fundamental level a brute-force molecular-dynamics approach does not attempt to take advantage of any inherent simplifying features present in the problem. For example, the method makes no use of the essentially harmonic character of the lattice. An attempt to formulate a collision theory which avoids some of these defects is sketched in the following section.

3.3 Generalized Langevin Treatment of Gas-Surface Dynamics

As discussed in the previous section a brute-force molecular-dynamics simulation of gas-surface dynamics, although simple in principle, is a large computational problem. Though these direct methods will continue to be of use, particularly in providing numerical benchmarks for the calibration of more approximate methods, it will prove useful to search for more efficient methods. The chief defect of direct methods when applied to gas-solid scattering is that the essentially harmonic character of the lattice is not fully exploited. We expect that the strong, direct interaction with the solid will involve a relatively small number of lattice atoms, probably in the 1-10 range. The remainder of the lattice (even if it is finite) thus serves as a harmonic heat bath for this "primary" lattice zone. Since the response of a collection of oscillators can be written as a function of the driving force, in this case the time-dependent forces supplied by the primary-lattice atoms, it is unnecessary to consider explicitly the motion of the background lattice. It becomes a generalized solvent with which the primary lattice interacts. As in the case of simple Brownian motion, the solvent's effects can be accounted for by adding friction and noise terms to the equations of motion for the system of primary interest. Thus the effects of the background lattice are included without resorting to an explicit study of its dynamics. Since the time scale of the background-lattice motion is comparable to that of the primary lattice, the character of the friction and noise terms for the present problem will not be as simple as for ordinary Brownian motion. However, the relevant characteristics of these terms will emerge as specific *lattice properties* which are thus independent of the nature of the incident species. Consequently these properties for any particular solid can be computed once and for all. This contrasts with direct molecular-dynamics methods which implicitly recompute these characteristics every time a collision trajectory is generated. By removing this inherent inefficiency it is hoped that the present approach can provide a formally more satisfying and a computationally more efficient approach to gas-surface dynamics.

In the present context it is important to emphasize that the above discussion applies equally well to situations where the original lattice is either finite or infinite. All that is required for the present analysis is that the background lattice be harmonic and that it be harmonically coupled to the primary zone. Obviously if the entire lattice consists of only a few atoms, direct molecular-dynamics simulations are appropriate. However, even submicron-size particles contain a sufficiently large number of constituent atoms to make the present approach advantageous.

The present section outlines the approach discussed above. Detailed developments are given elsewhere [3.36-40]. This approach is an outgrowth of previous work along these lines, principally by ZWANZIG [3.4] and by GOODMAN [3.1,5]. Recently SHUGARD et al. [3.41] have presented related but independent developments.

The equations of motion for a gas impinging on a harmonic lattice are

$$m_g \ddot{\underline{X}} = -\frac{\partial V}{\partial \underline{X}}(\underline{X},\underline{x}) \tag{3.49}$$

$$\ddot{\underline{x}} = -\ddot{\underline{\underline{\omega}}}^2 \underline{x} - \frac{\partial V}{\partial \underline{x}}(\underline{X},\underline{x}) \quad . \tag{3.50}$$

Here $\underline{X}$ denotes the gas coordinate and $\underline{x}$ the lattice displacements, the latter in some suitably chosen mass weighted units. We assume that the gas particle interacts directly only with a relatively small subset of $\underline{x}$, the primary lattice. It is thus advantageous to partition $\underline{x}$ into a primary zone, $\underline{x}_p$, and a background lattice, $\underline{x}_q$. Since the gas does not interact directly with $\underline{x}_q$, we write

$$V(\underline{X},\underline{x}) \sim V(\underline{X},\underline{x}_p,\underline{x}_q \equiv 0) \quad . \tag{3.51}$$

In partitioned form (3.49,50) become

$$m_g\ddot{\underline{X}} = -\frac{\partial V}{\partial \underline{X}} \tag{3.52}$$

$$\ddot{\underline{x}}_p = -\underline{\underline{\omega}}^2_{pp}\underline{x}_p - \underline{\underline{\omega}}^2_{pq}\underline{x}_q - \frac{\partial V}{\partial \underline{x}_p} \tag{3.53}$$

$$\ddot{\underline{x}}_q = -\underline{\underline{\omega}}^2_{qp}\underline{x}_p - \underline{\underline{\omega}}^2_{qq}\underline{x}_q \quad . \tag{3.54}$$

In (3.53,54) the original frequency matrix, $\underline{\underline{\omega}}^2$, has been partitioned into $\underline{\underline{\omega}}^2_{pp}$, $\underline{\underline{\omega}}^2_{pq}$, etc. Using standard Laplace transform techniques, we can eliminate the background-lattice variables from (3.53). We obtain

$$m_g\ddot{\underline{X}} = -\frac{\partial V}{\partial \underline{X}} \tag{3.55}$$

$$\ddot{\underline{x}}_p = -\underline{\underline{\omega}}^2_{pp}\underline{x}_p - \frac{\partial V}{\partial \underline{x}_p} + \int_0^t \underline{\underline{\Theta}}(t-t')\underline{x}_p(t')dt' + \underline{R}(t) \quad . \tag{3.56}$$

Here $\underline{\underline{\Theta}}(t)$ is given by [note the distinction between $\underline{\underline{\Theta}}(t)$ and $\underline{\underline{\theta}}(t)$]

$$\underline{\underline{\Theta}}(t) = \underline{\underline{\omega}}^2_{pq}\underline{\underline{\theta}}(t)\underline{\underline{\omega}}^2_{qp} \quad , \tag{3.57}$$

where $\underline{\underline{\theta}}(t)$ is the inverse Laplace transform of $\underline{\underline{\theta}}(z)$,

$$\underline{\underline{\theta}}(z) = (\underline{\underline{z}}^2 + \underline{\underline{\omega}}^2_{qq})^{-1} \quad . \tag{3.58}$$

The "noise" term in (3.56) represents the propagation of the initial positions and velocities of the background lattice. Specifically,

$$\underline{R}(t) = -\underline{\underline{\omega}}^2_{pq}\underline{x}^R_q(t) \quad - \underline{\underline{\omega}}^2_{pq}[\dot{\underline{\underline{\theta}}}(t)\underline{x}_q(0) + \theta(t)\dot{\underline{x}}_q(0)] \quad . \tag{3.59}$$

Here $\underline{x}^R_q(t)$ is the "free" solution of (3.54) if we set $\underline{x}_p = 0$. It thus represents the unperturbed motion of the background lattice for a situation where the atoms of the primary lattice are clamped in their equilibrium positions. For thermal conditions the initial positions and velocities of the (harmonic) background lattice are random, with a known Gaussian distribution. Thus $\underline{R}(t)$ is Gaussian random noise, a fact that will later prove useful. Equation (3.56) can be case in a velocity rather than a displacement form, a step that makes closer contact with familiar Langevin formulations. Defining $\underline{\underline{\beta}}(t)$ as

$$\underline{\underline{\beta}}(t) = \int_t^{\infty} \underline{\underline{\theta}}(t')dt' \quad , \tag{3.60}$$

(3.55,56) become

$$m_g\ddot{\underline{X}} = -\frac{\partial V}{\partial \underline{X}} \tag{3.61}$$

$$\ddot{\underline{x}}_p = -\underline{\underline{\Omega}}^2_{pp}\underline{x}_p - \frac{\partial V}{\partial \underline{x}_p} - \int_0^t \underline{\underline{\beta}}(t - t')\dot{\underline{x}}_p(t')dt + \underline{R}(t) - \underline{\underline{\beta}}(t)\underline{x}_p(0) \quad . \tag{3.62}$$

The effective frequency matrix, $\underline{\underline{\Omega}}^2_{pp}$, is given by

$$\underline{\underline{\Omega}}^2_{pp} = \underline{\underline{\omega}}^2_{pp} - \underline{\underline{\beta}}(0) \quad . \tag{3.63}$$

Since $\underline{\underline{\beta}}(t)$ decays to zero at large times, the $\underline{x}_p(0)$ term in (3.62) is a transient one.

Equations (3.61,62) [or equivalently (3.55,56)] express the gas-surface collision dynamics in generalized Langevin form. The principal advantage of this formulation is the elimination of explicit background-lattice dynamics. Effects of the background lattice are included, however, through the renormalized frequencies, and the friction and noise terms. From (3.62) we see that we can relax the assumption of a harmonic primary lattice. Only the existence of a harmonic background lattice coupled harmonically to the primary lattice is required to generate a result of the form in (3.62).

Since $\underline{R}(t)$ and $\underline{\underline{\beta}}(t)$ are related by [3.38]

$$\underline{\underline{\beta}}(t) = \langle \underline{R}(t)\underline{R}^T(0)\rangle/kT \quad , \tag{3.64}$$

we see that *all* background-lattice information relevant to the collision dynamics is summarized by $\underline{\underline{\beta}}(t)$. This quantity is a *lattice* property. Thus for any chosen system all necessary information can be computed once and for all *prior* to the dynamical study. This is convenient, particularly for the study of disordered systems. Techniques for the construction of these friction kernels and for the numerical simulations are discussed below.

Construction of the necessary Langevin friction kernels is most readily accomplished by identifying them as certain lattice correlation functions [3.42]. From (3.59) we see the "clamped-lattice" displacements, $\underline{x}^R_q(t)$, are given by

$$\underline{x}^R_q(t) = \dot{\underline{\underline{\theta}}}(t)\underline{x}_q(0) + \underline{\underline{\theta}}(t)\dot{\underline{x}}_q(0) \quad . \tag{3.65}$$

Thus

$$\underline{\underline{\theta}}(t) = \langle \underline{x}_q^R(t)\dot{\underline{x}}_q^T(0)\rangle / \langle \dot{\underline{x}}_q(0)\dot{\underline{x}}_q^T(0)\rangle$$
$$= -\langle \dot{\underline{x}}_q^R(t)\underline{x}_q^T(0)\rangle / kT \quad , \tag{3.66}$$

where T is the lattice temperature. Thus $\underline{\underline{\theta}}(t)$ and hence $\underline{\underline{\Theta}}(t)$ or $\underline{\underline{\beta}}(t)$ are available. As discussed by ADELMAN and DOLL [3.38] it is possible to avoid the construction of the large number of background-lattice correlation funcitions implicit in approaches based on (3.66). The equations of motion for the primary lattice in the absence of the gas [(3.56) with V = 0] can be solved yielding

$$\underline{x}_p(t) = \dot{\underline{\underline{\chi}}}(t)\underline{x}_p(0) + \underline{\underline{\chi}}(t)\dot{\underline{x}}_p(0) + \int_0^t {}_{=}(t - t')\underline{R}(t')dt' \tag{3.67}$$

where $\underline{\underline{\chi}}(t)$ is the inverse Laplace transform of

$$\underline{\underline{\chi}}(z) = [\underline{\underline{z}}^2 + \underline{\underline{\omega}}_{pp}^2 - \underline{\underline{\Theta}}(z)]^{-1} \quad . \tag{3.68}$$

Equation (3.67) implies

$$\underline{\underline{\chi}}(t) = - \langle \dot{\underline{x}}_p(t)\underline{x}_p^T(0)\rangle / kT \quad . \tag{3.69}$$

Thus by constructing only *primary*-lattice correlation functions we can construct $\underline{\underline{\chi}}(t)$ and hence $\underline{\underline{\chi}}(z)$. From (3.68) we see

$$\underline{\underline{\Theta}}(z) = \underline{\underline{z}}^2 + \underline{\underline{\omega}}_{pp}^2 - \underline{\underline{\chi}}^{-1}(z) \quad . \tag{3.70}$$

Once $\underline{\underline{\chi}}(z)$ is available from (3.69), inversion of (3.70) will yield $\underline{\underline{\Theta}}(t)$ and hence $\underline{\underline{\beta}}(t)$.

The essential point of the above discussion is that the necessary friction kernels are identifiable as certain lattice correlation functions. Once such an identification is made, the machinery for constructing such functions can be brought to bear on the problem of constructing the Langevin friction kernels. Possible approaches include direct dynamical matrix diagnolization [3.43,44],[4] mode-density approaches [3.41,45], moment methods [3.46] direct algebraic methods [3.42][5] and molecular-dynamics simulations.

Practical computational approaches have been developed for the numerical simulation of the generalized Langevin equations above [3.39,41,45]. The general

---

[4]See [3.43] for a general discussion of lattice dynamics.
[5]See [3.42] for a discussion of certain analytically soluble models.

features of these methods can be illustrated by considering the simple case of an ordinary Brownian particle [3.47]. If the particle is of mass m and the friction constant is β_0, the equations of motion for the particle's momentum, p, and position, x, are

$$\dot{p} = -\beta_0 p + R(t) \tag{3.71}$$

$$\dot{x} = p/m \quad . \tag{3.72}$$

R(t) in (3.71) is the random (white) noise supplied by the viscous medium. If we discretize time such that $t_n = n\Delta t$, where Δt is some small, specified time interval, we see that the values of p and x at t_{n+1} are related to those at t_n by

$$p_{n+1} = p_n + \int_{t_n}^{t_{n+1}} dt[-\beta_0 p(t)] + \int_{t_n}^{t_{n+1}} R(t)dt \tag{3.73}$$

$$x_{n+1} = x_n + \int_{t_n}^{t_{n+1}} [p(t)/m]dt \quad . \tag{3.74}$$

The integrals over the systematic parts of (3.73,74) can be evaluated by any number of standard quadrature methods. However, the simple Euler scheme will be sufficient to illustrate the essential result. Applying this method to (3.73,74) we obtain

$$p_{n+1} = p_n - \beta_0 p_n \Delta t + B_n(\Delta t) \tag{3.75}$$

$$x_{n+1} = x_n + p_n \Delta t/m \tag{3.76}$$

where

$$B_n(\Delta t) = \int_{t_n}^{t_{n+1}} R(t)dt \quad . \tag{3.77}$$

Since R(t) is Gaussian white noise, $B_n(\Delta t)$ is a Gaussian random variable with simple statistics. Specifically [3.47]

$$\langle B_n(\Delta t)\rangle = 0 \tag{3.78}$$

$$\langle B_n(\Delta t)B_{n'}(\Delta t)\rangle = (2m\beta_0 kT\Delta t)\delta_{nn'} \quad . \tag{3.79}$$

Thus, the Langevin equations (3.71,72) can be propagated much as any standard differential equations except that at each integration cycle a suitable random variable is included. More-general systems of equations involving several interacting Brownian particles can be handled in a similar fashion.

As can be seen above, the simulation of Brownian equations of motion is essentially no more difficult than the simulation of ordinary Hamiltonian equations of motion. Indeed, already developed classical trajectory algorithms can be readily adapted to the task. Previous work [3.40,45] has indicated that the effects of the background lattice can not be well represented by approximating the $\underline{\underline{\beta}}(t)$ term in (3.62) by a local friction expression. The time scales for primary- and background-lattice motion are not sufficiently dissimilar. An approach which retains the above simplicity, however, involves the replacement of the original *large-harmonic-background* lattice by a relatively *small* set of Brownian oscillators. The characteristics of these oscillators and of their coupling to the primary lattice can be chosen such that they optimally reproduce desired features of the original system, the primary-lattice response, $\underline{\underline{X}}(t)$, for example. Such an approach would amount to the replacement of (3.52-54) by

$$m_g\ddot{\underline{X}} = -\frac{\partial V}{\partial \underline{X}} \tag{3.80}$$

$$\ddot{\underline{x}}_p = -\underline{\underline{\omega}}_{pp}^2\underline{x}_p - \underline{\underline{\omega}}_{pB}^2\underline{x}_B - \frac{\partial V}{\partial \underline{x}} \tag{3.81}$$

$$\ddot{\underline{x}}_B = -\underline{\underline{\omega}}_{Bp}^2\underline{x}_p - \underline{\underline{\omega}}_{BB}^2\underline{x}_B - \underline{\underline{\beta}}_0\dot{\underline{x}}_B + \underline{R}_B(t) \quad , \tag{3.82}$$

where the Brownian bath quantities are denoted by the subscript B. Equations (3.80-82) are convenient in that compared with (3.52-54) they represent a significant reduction in the number of degrees of freedom. At the same time, numerical simulation of these equations is straightforward.

It should be borne in mind that for ordinary Brownian equations of motion Fokker-Planck methods [3.47] are often attractive computational possibilities. These methods are appealing since the fundamental equation of motion is for the phase-space distribution itself rather than for individual trajectories. The structure of the Fokker-Planck equation in effect carries out a number of averages that must otherwise be performed by generating suitable trajectory ensembles. A preliminary application of the Fokker-Planck method to gas-surface scattering has been made [3.37]. In this application it was assumed that the full phase-space distribution was Gaussian in character with time-dependent first and second moments. Consequently the Fokker-Planck equation produced a set of first-order differential equations for these moments [3.48].[6] Integration of these equations was essentially

[6]For a general treatment of stochastic differential equations see [3.48].

of the same order of difficulty as constructing a *single* classical trajectory.[7] To date no such application has been attempted for the Brownian bath model discussed above.

Although the Brownian bath model may emerge as the most generally useful computational procedure, it is important to develop techniques for the direct simulation of the basic generalized Langevin equations. Such simulations are more involved than those for simple Brownian systems in two ways. First, the friction term is nonlocal, involving an integral over the entire history of the particle. Second, the noise is no longer "white," but rather of general Gaussian character. Discussed below are methods for treating each of these aspects. The essential result is that the problem of numerical simulation of generalized Langevin systems can again be reduced to a form which can be treated with algorithms that are only slight modifications of existing molecular-dynamics programs.

The basic method utilized to treat the friction term in (3.62) can be illustrated by the following example. If we suppose for the moment that $\beta(t)$ in a simple one-dimensional generalized Langevin equation is given by

$$\beta(t) = A \exp(-t/\tau) \cos\omega t \quad , \tag{3.83}$$

then the friction integral we must treat is given by

$$I_c(t) = \int_0^t A \exp[-(t - t')/\tau] \cos[\omega(t - t')]\dot{x}(t')dt' \quad . \tag{3.84}$$

Defining a related integral, $I_s(t)$, by

$$I_s(t) = \int_0^t A \exp[-(t - t')/\tau]\sin[\omega(t - t')]\dot{x}(t')dt' \quad , \tag{3.85}$$

we note [3.46]

$$\dot{I}_c(t) = - I_c(t)/\tau - \omega I_s(t) + A\dot{x}(t) \tag{3.86}$$

$$\dot{I}_s = \omega I_c(t) - I_s(t)/\tau \quad , \tag{3.87}$$

where

$$I_c(0) = I_s(0) = 0 \quad . \tag{3.88}$$

I_c and I_s thus satisfy a set of coupled first-order equations which can be integrated along with the equations for the original dynamical variables. A similar

---

[7] It proved essential to completely couple the equations of motion for the first and second moments. See [3.37] for details.

development would also be possible if the friction kernel were composed of a sum of damped trigonometric terms, or, more generally, a sum of terms whose derivatives generate a closed set of functions. The number of auxiliary equations will increase with the number and complexity of terms included in the friction-kernel expansion. The efficiency of the Langevin method relative to an explicit study of the background lattice will thus depend on the difficulty involved in obtaining acceptable friction-kernel expansions.[8]

Several approaches are possible with regard to the treatment of the "noise" term in (3.62). One is to exploit the mechanical origin of this term, the free propagation of the initial positions and velocities of the background lattice cf (3.65) . A suitably large ensemble of such trajectories could be generated once for any chosen solid system and appropriate segments could be used in (3.62). The advantage of such an approach is that these same trajectories could be utilized in (3.66) to derive the necessary friction kernels. Furthermore, such an approach would automatically guarantee that the fluctuation-dissipation theorem was satisfied (3.64). Alternatively, one can avoid mechanical models and resort to direct construction of the Gaussian noise. We again consider for simplicity a one-dimensional problem. By analogy with (3.73), the integral over the noise that is required for the propagation of the generalized Langevin equation is

$$B_n = \int_{t_n}^{t_{n+1}} R(t)dt \quad . \tag{3.89}$$

From the fluctuation-dissipation theorem (3.64) it is not hard to show that the first and second moments of B_n are [3.39]

$$\langle B_n \rangle = 0$$

$$\langle B_n B_m \rangle = kT \left\{ \int_{t_m-t_n}^{t_{m+1}-t_n} \beta(s)(t_{m+1} - t_n - s)ds \right.$$

$$\left. + \int_{t_m-t_{n+1}}^{t_m-t_n} \beta(s)(t_{n+1} - t_m + s)ds \right\} \quad . \tag{3.90}$$

Thus unlike the Brownian case the B_n values for successive integration steps are *correlated*. The number of steps over which the correlation is significant is determined by the nature of the decay of the friction kernel. If it is assumed that

[8]Work by ADELMAN and GARRISON [3.45] gives some indication that relatively compact expansions are possible

$$\langle B_n B_m \rangle \sim 0 \ , \quad n - m \geq p \quad , \tag{3.91}$$

then it can be shown [3.39] that the $\{B_n\}$ values can be constructed by suitably combining uncorrelated Gaussian random variables. In particular, we can write

$$B_n = \sum_{i=1}^{p} c_i \, g_{i+n} \quad , \tag{3.92}$$

where $\{g_i\}$ corresponds to a list of uncorrelated Gaussian random variables (with zero mean and unit standard deviation). The coefficients in (3.92) satisfy

$$\langle B_p B_\ell \rangle = \sum_{i=1}^{\ell} c_i c_{i+p-\ell} \quad , \quad \ell = 1, p \quad . \tag{3.93}$$

Methods for the generation of these coefficients are described elsewhere. Construction of the integrals over the noise term by means of (3.92) is particularly convenient in the near Brownian limit, where only a few terms will be present.

3.4 Sample Results

Applications of the present method to gas-surface collisions have to date been exploratory in nature. The objectives have been to assess the background-lattice dependence of experimentally observed quantities, to examine the relative efficiency of the present approach, and to establish practical computational procedures. It is too early for definite conclusions, but available model studies do indicate the following:

1) Background-lattice effects can be appreciable, making some provision for their inclusion essential.
2) Computational algorithms necessary to implement the generalized Langevin approach can be developed.
3) The efficiency of the present method is significantly greater than that of a brute-force molecular-dynamics approach.

Several of these model studies are discussed below.

The first study using the present ideas examined the collinear-atom-Brownian-oscillator model [3.37]. Although this model will ultimately prove too simplistic, it is sufficient to gain some feeling for the background-lattice effect on energy transfer. Figure 3.6 shows the computed (fixed-energy) accommodation coefficient for He-W parameters as a function of the friction constant acting on the surface atom. We see that the energy-transfer efficiency is initially an increasing function of the friction constant. However, as the frictional effects become large (overdamped oscillator), this trend reverses. Utilizing the Debye model's estimate [3.38] of $\beta_0 = \pi\omega_D/6$, we see that the frictional effects produce approximately at 15 percent change in α_E.

Similar and even larger effects were seen in the study of a collinear-atom-generalized-Langevin-oscillator model [3.39]. In this study the friction kernel was taken as

$$\beta(t) = \tau^{-1}\beta_0 \exp(-t/\tau) \quad .^9 \tag{3.94}$$

The sensitivity of the computed accommodation coefficients for Ne-W and Ar-W to β_0 is shown in Fig.3.7. Again frictional effects are appreciable, especially for the Ar-W results. At first glance it might be assumed that the large increase in α_E with respect to β_0 for the Ar-W results was a result of an increase in the sticking probability. However, as is shown in Fig.3.8 the sticking probability actually decreases with increasing β_0. This study showed, however, that the residence time of particles that did stick was an increasing function of β_0. This effect, at least for small β_0 values, proved dominant. This study suggests that an adequate background-lattice treatment will be particularly important for condensation studies.

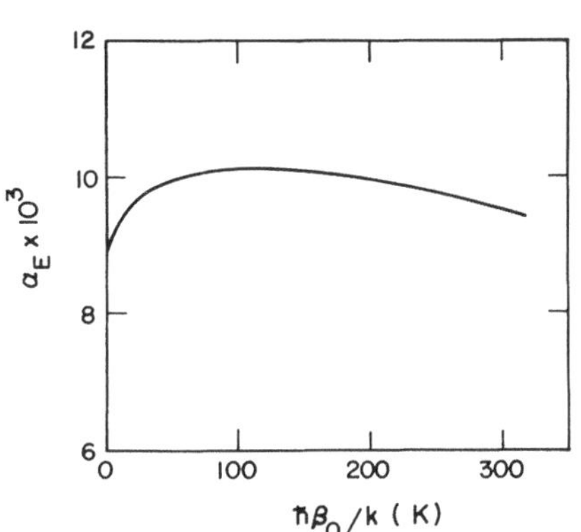

Fig.3.6. α_E as a function of the friction constant for the atom-Brownian-oscillator model [3.37]. Parameters are $T_g = 400$ K ($E_g \equiv kT_g/2$), $T_s = 100$ K, $\hbar\omega_0/k = 180$ K, $D/k = 65$ K, $\alpha = 1.05$ Å^{-1}, $m_g = 4$, and $m_s = 184$

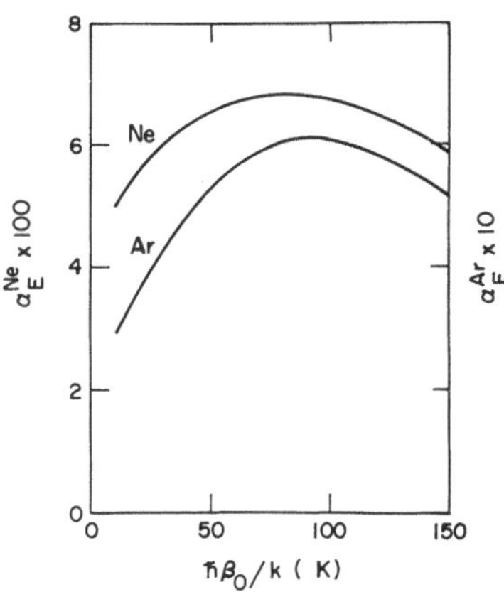

Fig.3.7. α_E as a function of β_0 for the atom-generalized-Langevin-oscillator model [3.39]. The friction kernel was of the form $\beta(t) = \tau^{-1}\beta_0 \exp(-t/\tau)$. A detailed parameter list is given in [3.39]

[9]The oscillator frequency was taken as constant [not (3.63)] in order to facilitate comparison with results in [3.37].

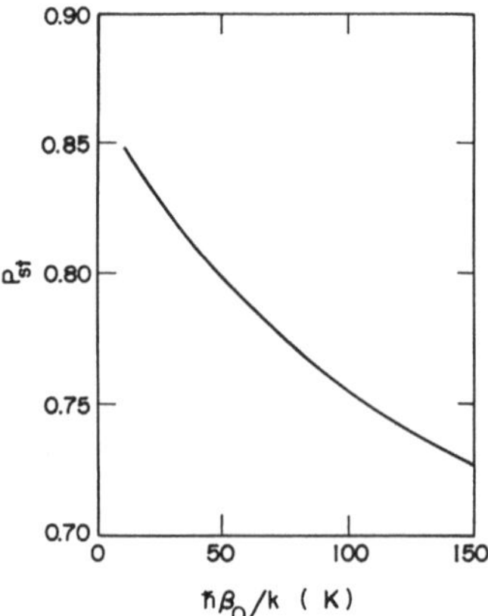

Fig.3.8. The sticking probability, P_{st}, or Ar-W as a function of β_0 for the model in Fig.3.7. "Sticking" is said to occur if the momentum perpendicular to the surface changes sign more than once during the collision

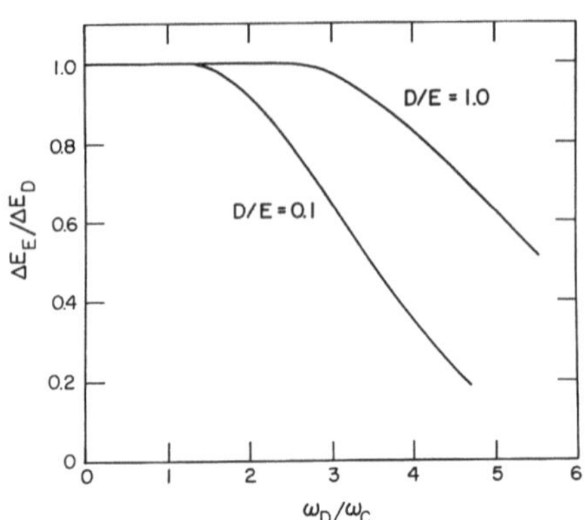

Fig.3.9. ΔE for a driven Einstein oscillator compared to that for a driven Debye model (3.29,38). The Einstein-oscillator frequency was taken as $\sqrt{0.6}\omega_D$. Other system parameters were ω_D = variable, m_g = 39.95, m_s = 108, D = 418 K, and α = 1.69 A^{-1}. Although obscured by the scale, both curves have a shallow minimum near ω_D/ω_C = 0 as a result of the characteristic shape of $F(\omega)$

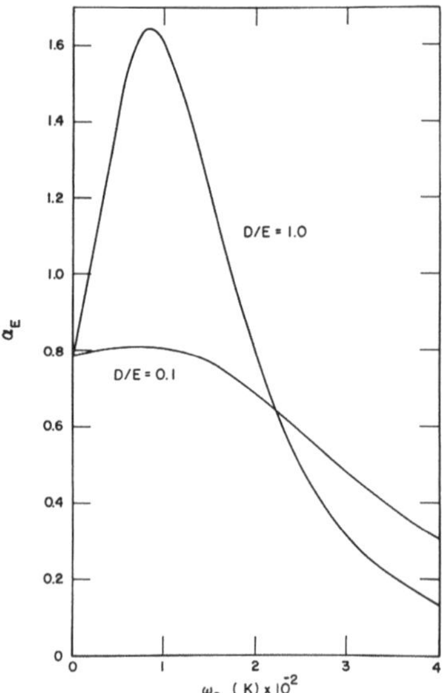

Fig.3.10. α_E for the driven Debye oscillator (3.38) as a function of ω_D. Parameters are those of Fig.3.9

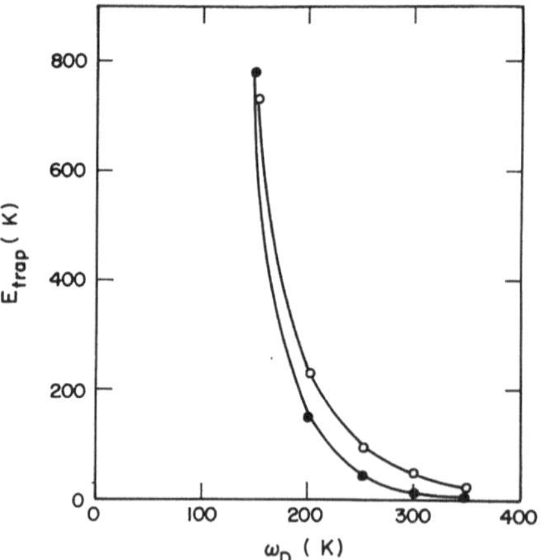

Fig.3.11. Trapping thresholds predicted by the driven Einstein (•) and driven Debye (o) models. Values in each case were determined by demanding α_E = 1 (3.29,38) Parameters are those of Fig.3.9

Figures 3.9-11 illustrate the importance of the background-lattice effects in a slightly different way. Shown are various results for the driven Einstein and Debye oscillator models of Sect.3.1. Figure 3.9 illustrates the ratio of the Einstein and Debye energy transfer (equivalently, the ratio of the Einstein and Debye accommodation coefficients) as a function of the adiabaticity parameter ω_D/ω_c. For abrupt collisions ($\omega_D/\omega_c < 1$) the impinging particles' disturbance does not have a chance to propagate any appreciable distance into the lattice. Consequently the Einstein model is adequate. However, for more languid collisions, appreciable energy can be transferred to the background lattice making the signal-oscillator model inadequate.[10] The actual dependence of α_E on the Debye frequency is illustrated in Fig.3.10. For larger D/E_g ratios we see that ${}_E$ reflects the characteristic shape of $F(\omega)$ (cf Fig.3.2). In this model α_E is not restricted to values less than unity. The quite different predictions given by the Einstein and Debye models concerning energy transfer for long duration collisions again emphasizes the importance of an adequate background-lattice treatment in condensation studies. For example, based on the results in Fig.3.9 we might expect the critical trapping energy would be sensitive to background-lattice effects. Shown in Fig.3.11 are the critical trapping energies predicted for the Ar-Ag system by the driven Einstein and Debye models. In each case the critical trapping energy was computed by finding the first energy for which $\alpha_E = 1$ as E_g is reduced from large values. We see that the trapping energy is very sensitive to the Debye frequency and to background-lattice effects. The results in Figs.3.9-11 are very similar to those reported by GARRISON and ADELMAN [3.49] on the basis of detailed numerical studies.

3.5 Conclusion

It appears that the generalized-Langevin-equation approach is a useful one for gas-surface dynamical studies. The approach provides a useful conceptual framework and offers inherent efficiency. The exact nature of the numerical details must await further studies. However, currently available model calculations suggest that the present method is a useful one.

---

[10]The general statement of this result is that in order for a collision model to be dynamically correct, the target's response for times of the order of the collision time must accurately approximate the response of the real system.

Acknowledgments

JDD would like to thank Dr. J.H. SULLIVAN and the Los Alamos Scientific Laboratory staff for the hospitality extended in a recent visit during which this work was completed. Partial financial support from the temporary Los Alamos NRIP program, the National Science Foundation, the Department of Energy, and from the Alfred P. Sloan Foundation is gratefully acknowledged.

References

3.1 F.O. Goodman, H.Y. Wachman: *Dynamics of Gas-Surface Scattering* (Academic Press, New York 1976)
3.2 W.H. Weinberg: Adv. Colloid Interface Sci. *4*, 301 (1975)
3.3 R.E. Stickney: *Adv. Atomic Mol. Phys.*, ed. by D.R. Bates, I. Estermann (Academic Press, New York 1967) p.143
3.4 R. Zwanzig: J. Chem. Phys. *32*, 1173 (1960)
3.5 F.O. Goodman: J. Phys. Chem. Solids *23*, 1269, 1491 (1962)
3.6 L. Landau, E. Teller: Phys. Z. Sowjetunion *11*, 18 (1936)
3.7 D.M. Gilbey: *Rarefield Gas Dynamics*, Vol.1, ed. by C.L. Brundin (Academic Press, New York 1967) p.121
3.8 F.E. Heidrich, K.R. Wilson, D. Rapp: J. Chem. Phys. *54*, 3885 (1971)
3.9 P.C. Martin: *Measurements and Correlation Functions* (Gordon and Breach, New York 1968)
3.10 R.D. Levine, R.B. Bernstein: *Molecular Reaction Dynamics* (Oxford University Press, New York 1974) Chap.5
3.11 S.A. Adelman, J.D. Doll: J. Chem. Phys. *64*, 2375 (1976)
3.12 See, for example, R.N. Porter: Annu. Rev. Phys. Chem. *25*, 317 (1974)
3.13 W.H. Miller: Adv. Chem. Phys. *25*, 69 (1974)
3.14 W. Eastes, J.D. Doll: J. Chem. Phys. *60*, 297 (1974)
3.15 R.V. Hogg, A.T. Craig: *Introduction to Mathematical Statistics*, 2nd ed. (Macmillan, New York 1965) Sect.4.3
3.16 D.W. Noid, M.L. Koszykowski, R.A. Marcus: J. Chem. Phys. *67*, 404 (1977)
3.17 R.B. Walker, R.K. Preston: J. Chem. Phys. *67*, 2017 (1977)
3.18 M.L. Koszykowski, D.W. Noid, J.D. McDonald, R.A. Marcus: J. Chem. Phys. (to be published)
3.19 J.D. McClure: J. Chem. Phys. *51*, 1687 (1969)
3.20 J.D. McClure: J. Chem. Phys. *52*, 2712 (1970)
3.21 J.D. McClure: J. Chem. Phys. *57*, 2810 (1972)
3.22 J.D. McClure: J. Chem. Phys. *57*, 2823 (1972)
3.23 R.A. Oman, A. Bogan, C.H. Li: *Rarefield Gas Dynamics*, Suppl.3, Vol.2, ed. by J.H. de Leeuw (Academic Press, New York 1966)
3.24 R.A. Oman: J. Chem. Phys. *48*, 3919 (1968)
3.25 R.A. Oman: *Rarefield Gas Dynamics*, Suppl.4, Vol.1, ed. by C.L. Brundin (Academic Press, New York 1967)
3.26 R.A. Oman: *Rarefield Gas Dynamics*, Suppl.5, Vol.2, ed. by L. Trilling, H.Y. Wackman (Academic Press, New York 1969)
3.27 L.M. Raff, J. Lorenzen, B.C. McCoy: J. Chem. Phys. *46*, 4265 (1967)
3.28 J.L. Lorenzen, L.M. Raff: J. Chem. Phys. *49*, 1165 (1968)
3.29 J.L. Lorenzen, L.M. Raff: J. Chem. Phys. *52*, 1133 (1970)
3.30 J.L. Lorenzen, L.M. Raff: J. Chem. Phys. *52*, 6134 (1970)
3.31 A. Gelb, M. Cardillo: Surf. Sci. *59*, 128 (1976)
3.32 A. Gelb, M. Cardillo: Surf. Sci. *64*, 197 (1977)
3.33 J.H. McCreery, G. Wolken, Jr.: J. Chem. Phys. *63*, 2340 (1975)
3.34 J.H. McCreery, G. Wolken, Jr.: J. Chem. Phys. *63*, 4072 (1975)
3.35 J.H. McCreery, G. Wolken, Jr.: J. Chem. Phys. *64*, 2845 (1976)
3.36 S.A. Adelman, J.D. Doll: J. Chem. Phys. *61*, 4242 (1974); *62*, 2518 (1975)

3.37 J.D. Doll, L.E. Myers, S.A. Adelman: J. Chem. Phys. *63*, 4908 (1975)
3.38 S.A. Adelman, J.D. Doll: J. Chem. Phys. *64*, 2375 (1976)
3.39 J.D. Doll, D.R. Dion: J. Chem. Phys. *65*, 3762 (1976)
3.40 S.A. Adelman, J.D. Doll: Acc. Chem. Res. *10*, 378 (1977)
3.41 M. Shugard, J.C. Tully, A. Nitzan: J. Chem. Phys. *66*, 2354 (1977)
3.42 See J.D. Doll, D.R. Dion: J. Chem. Phys. *67*, 3181 (1977)
3.43 A.A. Maradudin, E.W. Montroll, G.H. Weiss: *Theory of Lattice Dynamics in the Harmonic Approximation* (Academic Press, New York 1963)
3.44 F.E. de Wette, G.P. Alldredge: Methods Comput. Phys. *15*, 163 (1976)
3.45 S.A. Adelman, B. Garrison: J. Chem. Phys. *65*, 3751 (1976)
3.46 R.I. Cukier, J.C. Wheeler: J. Chem. Phys. *60*, 4639 (1974)
3.47 S. Chandrasekhar: Rev. Mod. Phys. *15*, 1 (1943)
3.48 N.G. van Kampen: Phys. Rep. *24*, 171 (1976)
3.49 B.J. Garrison, S.A. Adelman: Surf. Sci. *66*, 253 (1977)

4. Inelastic Light Scattering

P. J. McNulty, H. W. Chew, and M. Kerker

With 17 Figures

Inelastic light scattering processes such as Raman and fluorescent scattering have been utilized recently as diagnostic probes for small particles and biological cells. The model of inelastic scattering described in this chapter assumes that the active molecules within a spherical particle can be represented by a classical oscillating electric dipole. The pumping field at the incident frequency at the location of the molecule is given by classical elastic scattering theory. The far-field at the shifted frequency is obtained by solution of the boundary value problem of a radiating dipole located at an arbitrary position within a dielectric sphere. Calculations have been carried out to illustrate the effect of particle size, particle refractive index and the distribution of active molecules within the sphere upon the inelastically scattered radiation.

4.1 Overview

It has long been known that the elastic scattering of light or other forms of electromagnetic radiation by small particles is strongly dependent upon the size, shape, and refractive index of the particle as well as the wavelength of the light. It is less well appreciated that fluorescence and Raman scattering by active molecules embedded in or comprising small particles will also be strongly affected by the presence of the particle. The effect of the particle on the inelastic scattering will be quite pervasive. The angular distribution, intensity, polarization, and the emission spectra will all demonstrate some dependence upon the geometrical and optical properties of the particle as well as the spatial distribution of the active molecules within it.

One consequence of this dependence of particular significance to remote sensing of aerosols is that it would be a mistake to assume that any fluorescent or Raman signals are simply proportional to the number of active molecules contained in the particle.

Until recently light scattering by particles has been considered almost synonymous with the elastic scattering of photons by small particles. Most treatments are based at least conceptually on the Lorenz-Mie formulation of the elastic scattering of electromagnetic plane waves by a uniform dielectric sphere. This chapter is an attempt to outline a formalism through which light scattering theory can be extended to inelastic scattering and to demonstrate by illustrative calculations the substantial particle dependence effects. The treatment is simplified by the introduction of vector spherical harmonics. This involves some changes in notation from that normally used. For the reader's convenience the Lorenz-Mie model is briefy derived in Sect.4.2.1 using this notation. Some recent progress on the long-standing problems of scattering by nonspherical particles and particles having internal structure are also briefly discussed in Sects.4.2.2,3, respectively.

Inelastic scattering is considered in Sect.4.3. The analysis should be equally applicable to fluorescence and Raman scattering. In general the active molecules that participate in inelastic scattering may have preferred directions of polarization of the exciting radiation for optimum absorption and similarly fixed polarization directions for emission. These directions may or may not be randomly oriented within the particle. Moreover, the molecule may undergo a change in orientation between the absorption of the exciting photon and the emission of the scattered one. The molecular transitions may be electric-dipole or magnetic-dipole or higher-order transitions. Calculations using the model must, therefore, be quite specific and may in general be complicated. The emphasis in Sect.4.3 after describing the model will be on demonstrating the extent of the effect of the particle's geometrical and optical properties for the relatively simple case of active molecules that are isotropically polarizable and undergo electric dipole transitions. The case of coherent inelastic scattering is covered in Sect.4.4.

4.1.1 Background

The scattering of electromagnetic radiation by small particles has been a subject of scientific investigation in a variety of disciplines for over a century. This interest has until quite recently focused almost exclusively on the phenomenon of elastic scattering, i.e., where the incident and scattered photons have the same wavelength. For detailed treatment of the elastic scattering of photons by small particles the reader is referred to the standard references of VAN DE HULST [4.1], KERKER [4.2], and BORN and WOLF [4.3]. Most treatments of eleastic light scattering by particles are based at least in part on the Lorenz-Mie formalism in which the elastic scattering of light by uniform dielectric spheres is treated as a classical boundary value problem for the electromagnetic fields at the surface of the particle. Calculations based on it for elastic scattering by a single particle are limited in accuracy only by the difficulties of computation and the errors inherent in representing the particle by a homogeneous dielectric sphere. The computational difficulties are mitigated by the availability of large fast computers, and for many applications the particles or droplets are in fact quite spherical in shape and uniform in composition.

It is interesting that the early investigations of elastic light scattering resulted from the need to understand natural phenomena involving color effects including the blue color of the sky, the red color of sunset, the rainbow, and the brilliant colors of certain metal sols. These color effects were shown to result from the dependence of elastic scattering and absorption by small particles on the wavelength. Other phenomena that stimulated light scattering theory included the transhorizon propagation of radiowaves, the zodiacal light, the effects of the interstellar particles upon starlight, and the scattering of sunlight by aerosols, ice crystals, and water droplets in the atmosphere. More recently, cataract formation and its effects on vision have been interpreted in terms of light scattering [4.4].

In addition to explaining natural phenomena, light scattering may be applied as a diagnostic tool for determining the physical and optical properties of small particles, including biological cells [4.5]. Light scattering is nondestructive and capable of being applied to remote sensing. It has been a particularly fruitful technique in the study of particulate systems ranging from simple molecules to macromolecules to colloids. The development of lidar techniques has increased its potential for remote sensing, particularly in studies of the atmosphere [4.6].

However, elastic light scattering cannot yield information about particular molecular species contained in the particles of interest and in order to learn about these, there has been a growing interest in using inelastic scattering processes such as the Raman effect and fluorescence as diagnostic probes. Raman microprobes have been developed by DELHAYE and DHAMELINCOURT [4.7] and ETZ and ROSASCO [4.8] in which a laser source is focused onto a single particle of mass as little as 10^{-12}g, and microanalysis has been carried out for samples obtained from urban air, minerals, power-plant emissions, insecticide sprays, biological specimens, etc. [4.8]. Fluorescence is widely used in a variety of rapid-flow instruments in order to identify and separate categories of biological cells and also in order to follow cell development processes [4.5]. The biological cell is tagged with fluorescent dyes which attach to particular molecules such as DNA, RNA, phospholipids, etc., providing specific probes for these molecules and the structures in which these are contained.

There is in all of this work an underlying assumption that the physical effects occurring in macroscopic samples also apply when the active molecules are distributed within particles with dimensions comparable to the wavelength of the exciting radiation. ROSASCO et al. [4.9] have demonstrated that the Raman spectra of particles consisting of a broad range of substances are in agreement with the spectra of the same substances measured in the form of bulk samples, thereby permitting qualitative identification of particular molecular species.

However, quantitative analysis is more complicated. We [4.10,11] have recently given theoretical arguments to show that the angular distribution, intensity, and polarization of the Raman or fluorescent scattering signal will not only depend upon the number of active molecules but also upon the particle size, shape, refractive index, internal structure, and the distribution of the active molecules within the particle. This will be illustrated in this chapter by representative calculations.

4.2 Elastic Scattering

4.2.1 Scattering by a Uniform Dielectric Sphere (Lorenz-Mie Scattering)

Elastic scattering of electromagnetic waves by spherical particles has been extensively reviewed in the literature [4.1,2]. We will only summarize the results here in a notation which will prove convenient for the inelastic-scattering calculations to be discussed in Sect.4.3.

Consider a dielectric sphere (medium 1) of radius a embedded in an external medium (medium 2). The physical quantities pertaining to the two media will be labeled with suffixes 1 and 2 respectively. Let a circularly polarized plane wave of angular frequency ω_0 be incident on the sphere along the z-axis with electric and magnetic fields given by

$$\underline{E}_{inc} = (\hat{\varepsilon}_1 \pm i\hat{\varepsilon}_2)\exp(ik_2z) = \sum_{\ell,m} \{(ic/n_2^2\omega_0)\alpha_E(\ell,m)\nabla\times[j_\ell(k_2r)\underline{Y}_{\ell\ell m}(\hat{r})] + \alpha_M(\ell,m)j_\ell(k_2r)\underline{Y}_{\ell\ell m}(\hat{r})\} \quad , \tag{4.1}$$

$$\underline{B}_{inc} = \hat{\varepsilon}_3\times\underline{E}_{inc} = \mp i\underline{E}_{inc} = \sum_{\ell,m} \{\alpha_E(\ell,m)j_\ell(k_2r)\underline{Y}_{\ell\ell m}(\hat{r}) - (ic/\omega_0)\alpha_M(\ell,m)\nabla\times[j_\ell(k_2r)\underline{Y}_{\ell\ell m}(\hat{r})]\} \tag{4.2}$$

where $\hat{\varepsilon}_1$, $\hat{\varepsilon}_2$, $\hat{\varepsilon}_3$ denote unit vectors along the x, y, z axis, respectively and

$$\alpha_M(\ell,m) = i^\ell[4\pi(2\ell+1)]^{\frac{1}{2}}\delta_{m,1} \quad , \tag{4.3}$$

$$\alpha_E(\ell,m) = \mp i\alpha_M(\ell,m) \quad . \tag{4.4}$$

Our notation is the same as that of JACKSON [4.12] except for the vector spherical harmonics for which we use the notation of EDMONDS [4.13]. This will produce transmitted fields inside the particle

$$\underline{E}_t = \sum_{\ell,m} \{(ic/n_1^2\omega_0)\gamma_E(\ell,m)\nabla\times[j_\ell(k_1r)\underline{Y}_{\ell\ell m}(\hat{r})] + \gamma_M(\ell,m)j_\ell(k_1r)\underline{Y}_{\ell\ell m}(\hat{r})\} \quad , \tag{4.5}$$

$$\underline{B}_t = \sum_{\ell,m} \{\gamma_E(\ell,m) j_\ell(k_1r)\underline{Y}_{\ell\ell m}(\hat{r}) - (ic/\omega_0)\gamma_M(\ell,m)\nabla\times[j_\ell(k_1r)\underline{Y}_{\ell\ell m}(\hat{r})]\} \quad , \tag{4.6}$$

and scattered fields outside the particle

$$\underline{E}_{sc} = \sum_{\ell,m} \{(ic/n_2^2\omega_0)\beta_E(\ell,m)\nabla\times[h_\ell^{(1)}(k_2r)\underline{Y}_{\ell\ell m}(\hat{r})] + \beta_M(\ell,m)h_\ell^{(1)}(k_2r)\underline{Y}_{\ell\ell m}(\hat{r})\} \quad , \tag{4.7}$$

$$B_{sc} = \sum_{\ell,m} \{\beta_E(\ell,m)h_\ell^{(1)}(k_2r)\underline{Y}_{\ell\ell m}(\hat{r}) - i(c/\omega_0)\beta_M(\ell,m)\nabla\times[h_\ell^{(1)}(k_2r)\underline{Y}_{\ell\ell m}(\hat{r})] \quad . \tag{4.8}$$

The expansion coefficients for the transmitted and scattered fields are determined by the boundary conditions at $r = a$,

$$\hat{n}\times\underline{E}_t = \hat{n}\times(\underline{E}_{inc} + \underline{E}_{sc}) \tag{4.9}$$

and

$$\hat{n}\times\underline{H}_t = \hat{n}\times(\underline{H}_{inc} + \underline{H}_{sc}) \quad . \tag{4.10}$$

The results are

$$\gamma_E(\ell,m) = \frac{(in_1^2/\mu_2k_2a)\alpha_E(\ell,m)}{\varepsilon_1 j_\ell(k_1a)[k_2ah_\ell^{(1)}(k_2a)]' - \varepsilon_2 h_\ell^{(1)}(k_2a)[k_1aj_\ell(k_1(a)]'} \quad , \tag{4.11}$$

$$\gamma_M(\ell,m) = \frac{-i(\mu_1/k_2a)\alpha_M(\ell,m)}{\mu_2 h_\ell^{(1)}(k_2a)[k_1aj_\ell(k_1a)]' - \mu_1 j_\ell(k_1a)[k_2ah_\ell^{(1)}(k_2(a)]'} \quad ; \tag{4.12}$$

$$\beta_E(\ell,m) = \frac{\{\varepsilon_2 j_\ell(k_2a)[k_1aj_\ell(k_1a)]' - \varepsilon_1 j_\ell(k_1a)[k_2aj_\ell(k_2a)]'\}\alpha_E(\ell,m)}{\varepsilon_1 j_\ell(k_1a)[k_2ah_\ell^{(1)}(k_2a)]' - \varepsilon_2 h_\ell^{(1)}(k_2a)[k_1aj_\ell(k_1a)]'} \quad , \tag{4.13}$$

$$\beta_M(\ell,m) = \frac{\{\mu_1 j_\ell(k_1a)[k_2aj_\ell(k_2a)]' - \mu_2 j_\ell(k_2a)[k_1aj_\ell(k_1a)]'\}\alpha_M(\ell,m)}{\mu_2 h_\ell^{(1)}(k_2a)[k_1aj_\ell(k_1a)]' - \mu_1 j_\ell(k_1a)[k_2ah_\ell^{(1)}(k_2a)]'} \quad ; \tag{4.14}$$

where the properties of the Wronskians of the spherical Bessel functions have been used to simplify the numerators of the expansion coefficients of the transmitted fields. Note that the denominators for the expansion coefficients of the scattered fields are the same as the corresponding ones for the transmitted fields. Hence if a particular coefficient for the scattered field is large because of the near vanishing of the denominators (as occurs near a resonance), the corresponding amplitude for the internal (transmitted) field becomes enhanced as well.

The angular distribution for Lorenz-Mie scattering by a uniform dielectric sphere is illustrated in Fig.4.1 where the incident radiation of wavelength λ = 488 nm is scattered by a uniform dielectric sphere of radius a and refractive index, m = 1.5 relative to the external medium. Curves are drawn for various values of the sphere's size parameter ($\alpha = 2\pi a/\lambda$).

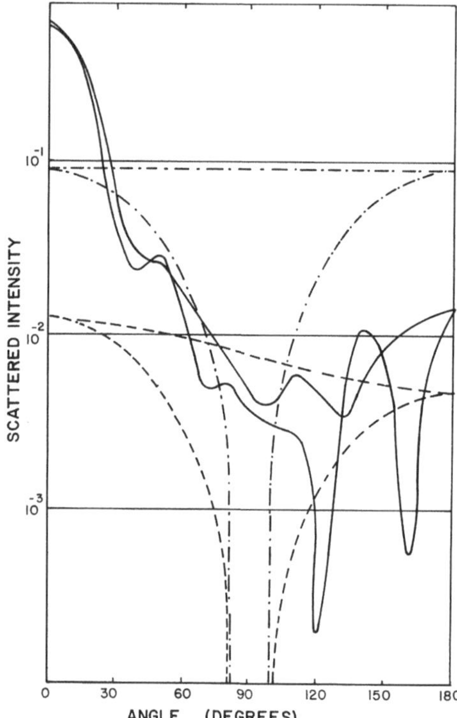

Fig.4.1. Lorenz-Mie scattering by a uniform dielectric sphere of refractive index 1.5 and radius a. Curves are drawn for values of the size parameter, $\alpha(2\pi a/\lambda)$ equal to 5 (——), 1 (----) and 0.01 (-·-·-·-)

4.2.2 Scattering by Spherical Particles with Internal Structure

Concentric Spheres

The solution to the scattering problem when the scatterer consists of concentric dielectric spheres has been worked out by ADEN and KERKER [4.14] and is reviewed in [4.2]. The scattering and internal-field coefficients may be expressed as ratios of determinants involving spherical Bessel functions. While no difficulty in principle is involved, computational complications increase rapidly with the number of layers. This problem has many interesting applications. For example, snow or ice particles falling in a relatively warm atmosphere may have a melted liquid layer surrounding the solid phase which would have significant effects on the scattering cross sections. Furthermore, by adjusting the geometrical and optical parameters of the concentric layers, it is possible to fabricate particles which have abnormally low scattering cross sections and which are, therefore, virtually invisible if the size of the particle is appreciably smaller than the wavelength [4.15,16]. Rough modeling of biological cells is also possible.

Approximate Methods

A great variety of approximate methods for electromagnetic scattering are available. The simplest of these, the Rayleigh-Gans approximation, is discussed in all standard texts [4.2]. Perturbation and variational methods are discussed, for example, in MORSE and FESHBACH [4.17] and ERMA [4.18]. Some of the more modern techniques are reviewed in MITRA [4.19] and USLENGHI [4.20]. In recent years, matrix methods have been developed and applied successfully to a number of scattering problems. Many of these are based on integral or matrix equations which are exact in principle, but which have to be truncated in some manner to obtain numerical results. In particular, the T-matrix approach as formulated by WATERMAN [4.21] and STRÖM [4.22] explicitly incorporates the symmetry of the scattering matrix due to the time-reversal invariance, and is sufficiently general to allow variations which yield different approximation schemes. An important feature of these methods is their applicability to scattering by objects of arbitrary shape. For some alternative formulations and numerical results see BARBER and YEH [4.23] and WATERMAN [4.24].

We have recently considered another approach, formulated by SAXON [4.25], and described in NEWTON [4.26], which is based on the fact that the scattered field may be expressed as an integral involving the internal field over the scatterer. For simplicity we assume the scatterer to be nonmagnetic. If $\underline{E}_{sc}(\hat{\underline{r}})$ denotes the scattered electric field at a point $\underline{r}$ far from the scatterer and $\underline{E}_{int}(\hat{\underline{r}}')$ denotes the electric field at a point $\underline{r}'$ inside the scatterer, we have

$$\underline{E}_{sc}(\underline{r}) = -\frac{k^2}{4\pi}\frac{\exp(ikr)}{r}\int[\varepsilon(\underline{r}') - 1][\underline{E}_{int}(\underline{r}') - \hat{r}\cdot\underline{E}_{int}(\underline{r}')\hat{r}]d^3r' \tag{4.15}$$

where $\varepsilon(\underline{r}')$ denotes the dielectric constant at the coordinate $\underline{r}'$ and the integration is carried over the scatterer only. Now the internal field satisfies a singular integral equation [4.27]

$$\begin{aligned}\underline{E}_{int}(\underline{r}) = \underline{E}_{inc}(\underline{r}) &+ k^2\int[\varepsilon(\underline{r}') - 1]G(\underline{r},\underline{r}')\underline{E}_{int}(\underline{r}')d^3r' \\ &- \int\{\nabla'[\nabla'\cdot\underline{E}(\underline{r}')]\}G(\underline{r},\underline{r}')d^3r'\end{aligned} \tag{4.16}$$

where k denotes the wave number outside, $\underline{E}_{inc}(\underline{r})$ denotes the incident field (usually a plane wave) evaluated at the point $\underline{r}$,

$$G(\underline{r},\underline{r}') = \frac{\exp(ik)|\underline{r}-\underline{r}'|}{4\pi|\underline{r}-\underline{r}'|} \quad , \tag{4.17}$$

and the integration is again over the scatterer only. Thus if the integral equation can be solved by one of the known methods, the internal field may be used to compute the scattered field. In fact, even without solving the integral equation (4.17), intelligent guesses of the internal field often produce satisfactory results when inserted into (4.1) to calculate $\underline{E}_{sc}(\underline{r})$ [4.28].

A major advantage of this method is that there is no restriction on the shape of the particle or its internal structure, although of course the size of the particle will be reflected in the amount of computer time needed to solve the integral equation (4.16).

4.2.3 Scattering by Nonspherical Particles

The number of scattering problems that can be solved analytically is severly limited by the inseparability of the vector wave equation in all but a very few coordinate systems. In the majority of cases various approximate methods have to be used. An excellent review of the analytic results for perfectly conducting bodies has been given by BOWMAN et al. [4.29]. These include circular, elliptic, parabolic, and hyperbolic cylinders; the wedge, the half plane, and other geometries. For infinite dielectric circular cylinders, see the review in KERKER [4.2].

Significant progress has been made recently in the scattering problem for a spheroidal particle of revolution by ASANO and YAMAMOTO [4.30]. In spite of this and other meritorious work [4.29,31], however, computational difficulties remain severe because of the lack of orthogonality of some of the eigenfunctions involved and the necessity of determination of eigenvalues by computing infinite determinants or continued fractions. These difficulties are aggravated when the radius exceeds appreciably the wavelength of the incident light. Nevertheless, with the rapid rate of progress in computer technology, solution of this problem should become increasingly accessible. Because of the availability of two size parameters, the spheroidal particle can be used to approximate a great variety of geometries and is, therefore, of considerable interest.

4.3 Incoherent Inelastic Scattering

In the previous section the elastic scattering by a spherical particle was treated classically. Of course, ultimately the photons incident on the particle are scattered by the molecules that comprise the particle. The scattered photons remain linked in phase and interfere coherently with the incident radiation and that scattered by the other molecules. The standard classical theory of elastic light scattering described in Sect.4.2 treats the process as a boundary value problem.

In the inelastic scattering considered here the scattered radiation arises from molecular transitions and has a frequency (and wavelength) that is different from the incident radiation. It is assumed in what follows that the inelastically scattered radiation can be measured independently of the incident and elastically scattered components at the exciting frequency by means of filters or a monochromator.

The inelastically scattered photons may or may not be scattered in phase. If they retain a phase relationship with one another the scattered radiation is coherent. If the inelastically scattered photons are emitted with random phases, as in spontaneous Raman and fluorescence, the radiation is said to be incoherently scattered. Because incoherent scattering seems to have the most immediate potential for applications in aerosol studies it will be emphasized in what follows. However, the case of coherent scattering by molecules embedded in particles is also of potential significance and will also be considered briefly in the next section.

Just as in the elastic case, the preferred method of treatment is to convert the scattering problem to a boundary value problem. In incoherent scattering the process reduces to a sequence in which individual molecules absorb light at the incident angular frequency ω_0, and then reemit light at the scattered angular frequency ω. If the molecules were in a gas or randomly distributed in a bulk medium the radiation pattern from each such active molecule would be identical, and by superposition the total radiation pattern would exhibit the same angular distribution and polarization. The signal strength would be proportional to the number of active molecules. However, the presence of the particle affects the radiation pattern of each active molecule. Even if the particle comprises only active molecules, the geometrical and optical properties of the particle affect each absorption and reemission.

Before considering the effects of the particle on the inelastic scattering by a molecule, let us consider the radiation pattern to be assumed for the active molecule. In order to emphasize, in this section, the effects of the optical and geometrical properties of the particle, a simple model is assumed in which the active molecule emits the photons through electric-dipole transitions and the molecule is assumed to be isotropically polarizable, i.e., its dipole moment $\underline{p}$ is given by

$$\underline{p} = \alpha' \underline{E}(\underline{r}') \tag{4.18}$$

where α' is a scalar and $\underline{E}(\underline{r}')$ is the local exciting field at the location of the active molecule. The model can be extended to magnetic-dipole and higher-order transitions, and to the case when the polarizability α in (4.18) is a tensor. However, the case of isotropic polarizability is sufficient to illustrate the effects the particle has on the radiation pattern.

4.3.1 Formalism for a Single Dipole

Consider a classical electric dipole of moment $\underline{p}$ oscillating with angular frequency at position $\underline{r}'$ in a bulk medium of refractive index n. The spatial part of the vector potential $\underline{A}_{Dip}$ produced at position $\underline{r}$ is given by

$$\underline{A}_{Dip}(\underline{r}) = -ik\underline{p}\, \exp(ik|\underline{r} - \underline{r}'|)/|\underline{r} - \underline{r}'|$$

$$= 4\pi k^2 \underline{p} \sum_{\ell,m} j_\ell(kr')h_\ell^{(1)}(kr)Y_{\ell m}(\hat{r}')Y^*_{\ell m}(\hat{r}') \tag{4.19}$$

where we have used the addition theorem for spherical harmonics. The magnetic field associated with this dipole for distance $r > r'$ is given by

$$\underline{B}_{Dip}(r) = \nabla x \underline{A}_{Dip} = 4\pi k^3 \sum_{\ell,m} j_\ell(kr')h_\ell^{(1)'}(kr)\hat{r}x\underline{p}Y^*_{\ell m}(\hat{r}')Y_{m}(\hat{r})$$

$$+ \frac{4\pi ik^2}{r} \sum_{\ell,m} j_\ell(kr')h_\ell^{(1)}(kr)Y^*_{\ell m}(\hat{r}')\underline{p}x[\hat{r}x\underline{L}Y_{\ell m}(\hat{r})] \tag{4.20}$$

where

$$h_\ell^{(1)'}(x) = dh_\ell^{(1)}(x)/dx \quad , \tag{4.21}$$

and $\underline{L}$ represents the operator

$$\underline{L} = -i\underline{r}x\nabla \quad . \tag{4.22}$$

The electric field of the electric dipole is then given by

$$\underline{E}_{Dip}(r) = \frac{1}{n}\underline{B}_{Dip}(\underline{r})x\hat{r} \quad . \tag{4.23}$$

The time-averaged power radiated per unit solid angle is

$$dP/d\Omega = cr^2|\underline{B}_{Dip}(\underline{r})|^2/(8\pi n) \quad . \tag{4.24}$$

The angular distribution of the power radiated by a dipole located at the origin for a distant observer is given in Fig.4.2. The curve labeled H represents the radiation pattern at large distances in the plane in which the dipole is oscillating. The minimum occurs in the direction along which the dipole is oscillating. The curve labeled V represents the radiation pattern obtained in a plane containing the dipole which is perpendicular to the direction of oscillation. In such a plane the radiation pattern is isotropic. In what follows it will be shown that the radiation patterns for dipoles embedded in small particles are significantly changed from that shown in Fig.4.2 except for locations near the center of the particle.

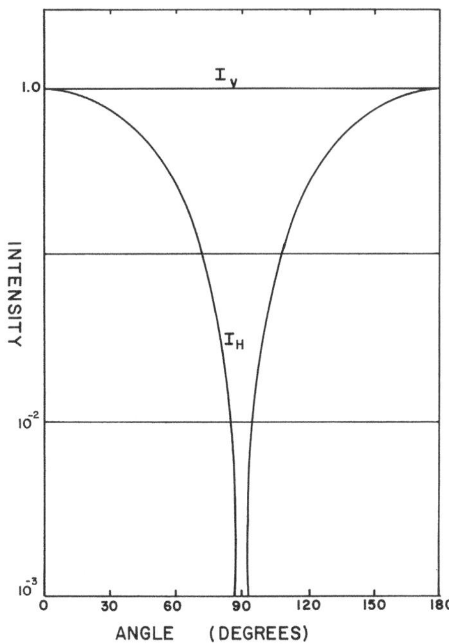

Fig.4.2. Radiation pattern for a classic dipole. The curves labeled I_H and I_V represent the respective radiation pattern at large distances in the plane and perpendicular to the plane in which the dipole oscillates

The angular distribution for inelastic scattering through electric-dipole transitions embedded in a bulk medium is the same when observed at large distances as that shown in Fig.4.2 as long as the molecules are isotropically polarizable and retain the orientation of their dipole moments between the absorption of the exciting photon and the subsequent emission. In that case, curve H represents scattering when the exciting light on the molecules is polarized in the scattering plane, and curve V represents polarization perpendicular to the scattering plane.

Dipole in a Small Uniform Dielectric Sphere

When a dipole is embedded within a particle whose size is comparable to the wavelength, the angular distribution of the emitted radiation may change significantly from that shown in Fig.4.2. Although the dipole may be embedded within a medium of the same composition as the bulk, the scattered radiation now depends upon the position of the dipole within the particle and upon the geometrical and optical properties of the particle. This occurs for two reasons. The strength of the field at the exciting frequency, which determines the induced dipole moment, varies sensitively with the location of each dipole within the particle. Moreover, there is an additional induced field at the shifted frequency which is necessary to satisfy the boundary conditions.

Figure 4.3 shows schematically a dipole at an arbitrary location $\underline{r}'$ within a dielectric sphere centered at the origin. The sphere has radius a and refractive index n_1. Again primed coordinates refer to the locations of the active molecules

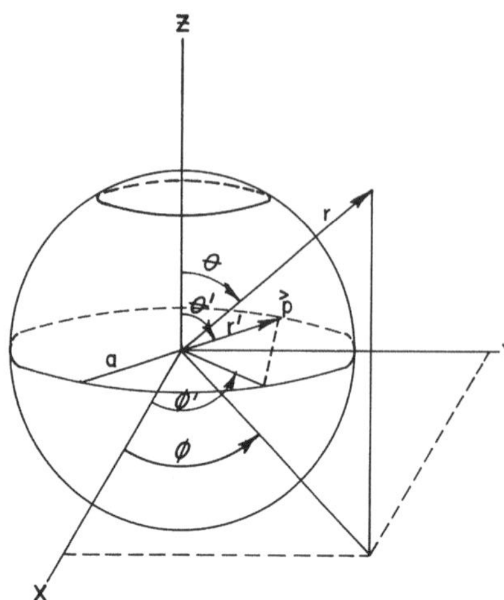

Fig.4.3. Schematic drawing of a spherical dielectric particle showing the dipole location within it. The exciting radiation is incident upon the particle along the +z direction

within the particle and the unprimed coordinates refer to the location of the observer at large distances outside the particle. The dipole is assumed to oscillate with angular frequency ω.

The field with angular frequency ω inside the particle includes contributions from the dipole and the particle

$$\underline{E}_1(r) = \underline{E}_{Dip}(\underline{r}) + \sum_{\ell,m} \{(ic/n_1^2\omega)b_E(\ell,m)\nabla\times[j_\ell(k_1r)\underline{Y}_{\ell\ell m}(\hat{r})] + b_M(\ell,m)j_\ell(k_1r)Y_{\ell\ell m}(\hat{r}) \tag{4.25}$$

$$\underline{B}_1(\underline{r}) = \underline{B}_{Dip}(\underline{r}) + \sum_{\ell,m} \{b_E(\ell,m)j_\ell(k_1r)\underline{Y}_{\ell\ell m}(\hat{r}) - (ic/\omega)b_M(\ell,m)\nabla\times[j_\ell(k_1r)\underline{Y}_{\ell\ell m}(\hat{r})]\} \tag{4.26}$$

where the dipole fields $\underline{E}_{Dip}(\underline{r})$ and $\underline{B}_{Dip}(\underline{r})$ are given by (4.20-23).

The contributions from the particle (4.25,26) are necessary to satisfy the boundary conditions given by (4.9,10) which state that the tangential components of $\underline{E}$ and $\underline{H}$ must be continuous across the boundary. The presence of fields at other frequencies inside or outside the particle can be ignored as long as suitable filters or monochromators are employed in measuring the scattered radiation.

The fields outside the particle can be expressed as

$$\underline{E}_2(\underline{r}) = \sum_{\ell,m} \{(ic/n_2^2\omega)C_E(\ell,m)\nabla\times[h_\ell^{(1)}(k_2r)\underline{Y}_{\ell\ell m}(\hat{r})] + C_M(\ell,m)h_\ell^{(1)}(k_2r)\underline{Y}_{\ell\ell m}(\hat{r})\} \tag{4.27}$$

$$\underline{B}_2(\underline{r}) = \sum_{\ell,m} \{C_E(\ell,m)h_\ell^{(1)}(k_2r)\underline{Y}_{\ell\ell m}(\hat{r})$$

$$-(ic/\omega)C_M(\ell,m)\nabla\times[h_\ell^{(1)}(k_2r)\underline{Y}_{\ell\ell m}(\hat{r})]\} \tag{4.28}$$

where n_2 and k_2 are the refractive index and wave number corresponding to the medium outside the particle.

In order to match the fields at the boundary the dipole fields of (4.20,23) must be put into suitable form

$$\underline{E}_{Dip}(r) = \sum_{\ell,m} \{(ic/n_1^2\omega)a_E(\ell,m)\nabla\times[h_\ell^{(1)}(k_1r)\underline{Y}_{\ell\ell m}(\hat{r})]$$

$$+ a_M(\ell,m)h_\ell^{(1)}(k_1r)\underline{Y}_{\ell\ell m}(\hat{r})\} \tag{4.29}$$

$$\underline{B}_{Dip}(r) = \sum_{\ell,m} \{a_E(\ell,m)h_\ell^{(1)}(k_1r)\underline{Y}_{\ell\ell m}(\hat{r})$$

$$-(ic/\omega)a_M(\ell,m)\nabla\times[h_\ell^{(1)}(k_1r)\underline{Y}_{\ell\ell m}(\hat{r})]\} \quad . \tag{4.30}$$

We have carried this out [4.10] and found that the expansion coefficients may be written

$$a_E(\ell,m) = [4\pi ik_1^3/\sqrt{2\ell + 1}]\underline{p} \cdot [\sqrt{\ell}j_{\ell+1}(k_1r')\underline{Y}^*_{\ell,\ell+1,m}(\hat{r}')$$

$$- \sqrt{\ell + 1}j_{\ell-1}(k_1r')\underline{Y}^*_{\ell,\ell-1,m}(\hat{r}')] \quad , \tag{4.31}$$

and

$$a_M(\ell,m) = 4\pi i(k_1^2\omega/c)j_\ell(kr')\underline{p} \cdot \underline{Y}^*_{\ell\ell m}(\hat{r}') \quad . \tag{4.32}$$

By matching the appropriate components of the fields at the boundary $r = a$, we found the following relationship between the expansion coefficients for the inelastically scattered field and $a_E(\ell,m)$ and $a_M(\ell,m)$

$$C_E(\ell,m) = f_\ell a_E(\ell,m)$$

$$C_M(\ell,m) = g_\ell a_M(\ell,m) \tag{4.33}$$

where

$$f_\ell = \frac{(i/k_1a)}{\{n_1^2j_\ell(k_1a)[k_2ah_\ell^{(1)}(k_2a)]'-h_\ell^{(1)}(k_2a)[k_1aj_\ell(k_1a)]'\}} \tag{4.34}$$

$$g_\ell = \frac{(i/k_1a)}{\{j_\ell(k_1a)[k_2ah_\ell^{(1)}(k_2a)]'-h_\ell^{(1)}(k_2a)[k_1aj_\ell(k_1a)]'\}} \quad . \tag{4.35}$$

The relative intensity of the radiation fields at the location of an observer outside the particle is given in terms of the dipole moment $\underline{p}$ as a function of angle by (4.24,30-35). Of course, in fluorescence or Raman scattering the relation between the dipole moment $\underline{p}$ in (4.31,32) and the exciting field strength may be complicated. If we assume for simplicity that the molecule is isotropically polarizable and ignore the absorption of photons of the localized exciting field, the dipole moment is given by (4.18). Computer programs have been developed [4.32] for calculating the radiation patterns for isotropically polarizable molecules using (4.18,24,30-35). The illustrative calculations that follow are based on these programs.

Effect of Dipole Location on Radiation Pattern

The effect that the particle has on the fluorescence or Raman scattering by molecules embedded within it can be strikingly illustrated by considering a single dipole. This dipole represents a microscopic aggregate of active molecules embedded within a homogeneous dielectric sphere. The angular distribution for the inelastically scattered intensity is plotted in Fig.4.4 for a single dipole at two positions along the x-axis. The x-axis lies in the scattering plane, and the exciting fields are incident on the particle along the +z direction in Fig.4.3. When the dipole is at the center the radiation pattern is essentially the same as that shown in Fig.4.2 for a dipole at the origin in a bulk medium of the same refractive index. For the example shown in Fig.4.4, $n_1 = 1.5$, $n_2 = 1.0$, $\omega_0 = 1.5\omega$ and the size parameter

$$\alpha = 2\pi a/\lambda \tag{4.36}$$

is $\alpha = 5.0$. Moving the dipole out along the x-axis introduces structure into the curves for both incident polarizations and shifts the position of the minima for horizontal incident polarization.

The effect of moving along the x-axis is shown in more detail in Fig.4.5 which plots the intensity at scattering angles of 30° and 180° against the dipole's location along the x-axis. The symmetry of the electric field for positive and negative values of x is mirrored by a corresponding symmetry in the radiation scattered at 180°. At 30° this symmetry is no longer preserved. The variation in scattered intensity is due to the variation of the intensity of the exciting field within the particle and the effects of the particle. For comparison the intensity of the local exciting field at the location of the dipole as given by the Lorenz-Mie theory [see (4.2)] is plotted as a dotted curve in Fig.4.5. The changes in the

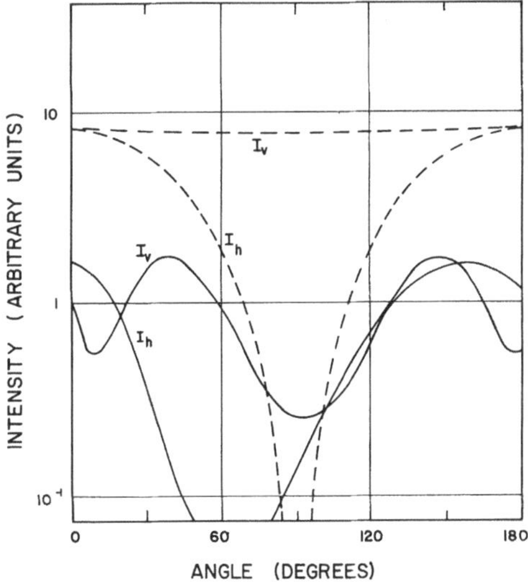

Fig.4.4. Inelastically scattered intensity vs scattering angle for two dipole locations along the x axis; α = 5, m_1 = 1.5, λ_0 = 488 nm, and λ = 732 nm. Curves are drawn for the cases of the incident light linearly polarized in the scattering plane, I_h, and perpendicular to it, I_v. The dashed and solid curves represent the scattering dipole located at x = 0.01a, y = z = 0 and x = 0.7a, y = z = 0, respectively

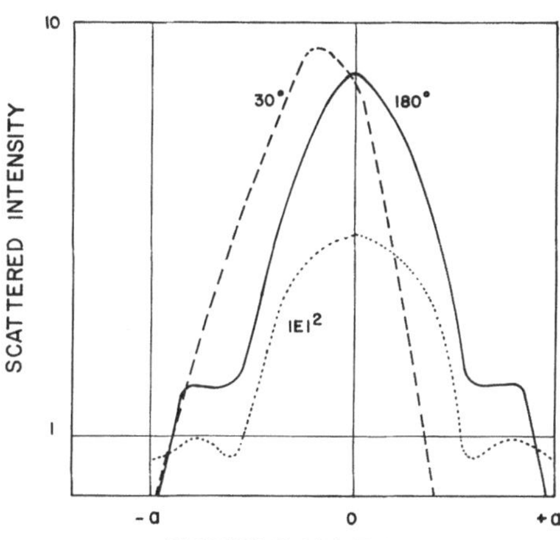

Fig.4.5. Inelastically scattered intensity I_h at scattering angles 30° and 180° as a function of position of the active molecule along the x axis; α = 5, m_1 = 1.5, λ = 1.5 λ_0. The power associated with the internal exciting field at λ_0 is plotted as a function of the molecule's location as a dotted curve

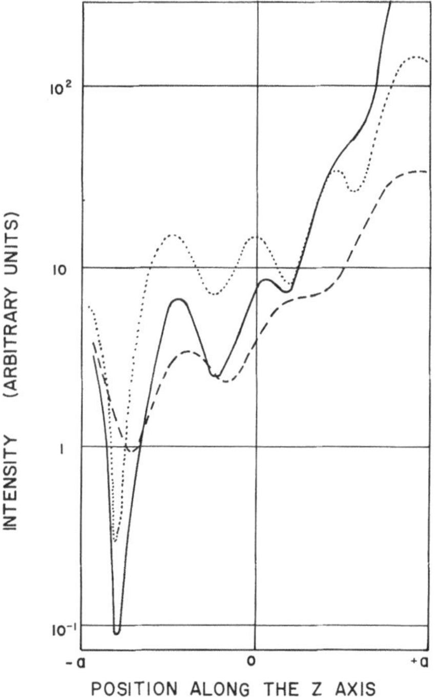

Fig.4.6. Inelastically scattered intensity I_h at scattering angle 180° as a function of active molecule location along the z axis; α = 5, m_1 = 1.5, λ = 1.5 λ_0. The power associated with the internal exciting field at each site is given by a dotted curve and the dashed curve represents I_h at 180° when the internal exciting field is replaced by the unperturbed incident field

fluorescence observed at 30° and 180° are apparently only partially due to the variation of the exciting field within the particle.

The most intense radiation pattern at 180° was obtained in Fig.4.4 for dipole locations along the x-axis when the dipole was near the center of the sphere. This is not true in general. For example, Fig.4.6 shows how the 180° signal varies with dipole location along the z-axis. In that case the signal from a dipole located near the front edge of the particle can be almost four orders of magnitude stronger than for positions near the back edge. The intensity of the exciting field at the location of the dipole is represented by the dotted curve in Fig.4.6 and it varies over two orders of magnitude in a manner similar to the 180° signal. The dielectric particle affects the signal at 180° in other ways. The dashed curve in Fig.4.6 represents the inelastically scattered signal at 180° when the internal exciting field is replaced by a uniform field. Even when the exciting field is a constant over the particle there is sufficient variation in the 180° signal as the dipole moves along the z-axis to explain the remaining two orders of magnitude. Thus, the effects due to the secondary field and the boundary conditions at the shifted frequency can be major contributions to the fluorescence or Raman scattering as well as the variation in the local exciting field intensity.

4.3.2 Incoherent Scattering by a Distribution of Dipoles

If the particle contains a number of dipoles and the scattering is incoherent (e.g., spontaneous Raman and fluorescence) the intensity of the light at the scattered frequency emerging from the particle is the sum of the contributions from each of the dipoles. The contribution from each dipole is different because the strength and direction of the exciting field varies with the location of the dipole with the particle. Moreover, the contribution from the dielectric medium of the particle necessary to satisfy the boundary conditions also depends upon the location of each dipole.

As other dipoles are added at various locations within the particle some of the effects observed in Figs.4.4-6 begin to diminish. This dependence can be illustrated by assuming a given number of dipoles are uniformly distributed over a spherical shell of radius $\underline{r}$ concentric with the spherical particle. The radiation pattern from an extended distribution of active molecules within a particle exhibits a clear dependence upon the distance from the molecules to the particle's center. In Fig.4.7 the angular distribution of the scattered radiation is plotted for a number of values of the radius. In this and the examples to follow, the results have been normalized to the same number of dipoles. As the radius increases the minimum at 90° fills in, and pronounced dissymmetry in the forward to backward scattering is introduced. Interestingly, except for scattering in the backward direction (and around 90° for the case of horizontal incident polarization) the inelastic scattering is greatest in Fig.4.7 when the dipoles are located near

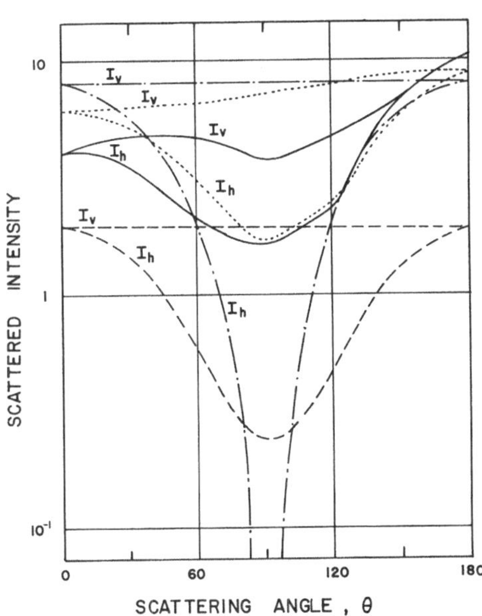

Fig.4.7. Inelastic scattered intensities I_h and I_v for arrays of active molecules uniformly distributed over spherical surfaces at radial distances a, 0.166a (----), 0.5a (····) and for molecules concentrated at the origin, (-.-.-.-) $\alpha = 5$, $m_1 = 1.5$, $\lambda = 1.5\lambda_0$. All curves are given for the same number of active molecules

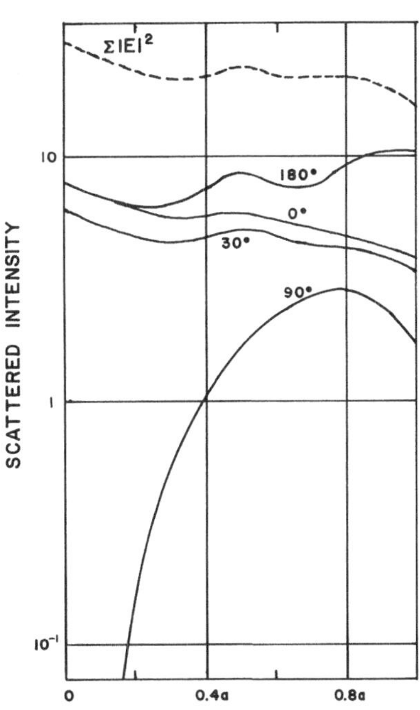

Fig.4.8. Inelastically scattered intensity I_h at scattering angles 0°, 30°, 90° and 180° as a function of radial distance of the array of active molecules uniformly distributed over a spherical surface $\alpha = 5$, $m_1 = 1{,}5$, $\lambda = 1.5\lambda_0$. The dashed curve represents the power of the exciting field averaged over the molecules locations

the center of the particle. The signal at 90° is particularly sensitive to the radius of the spherical shell. Figure 4.8 shows the relative intensity per dipole scattered through various angles of horizontally polarized incident radiation versus the radius of a dipole shell. The signal at 90° rises sharply in Fig.4.8 as the dipoles move farther from the origin even though the intensity of the exciting radiation summed over the individual dipole locations is simultaneously decreasing.

Uniformly Filled Sphere

The dependence of the radiation pattern from dipoles uniformly distributed within a spherical particle of relative refractive index $m_1 = n_1/n_2 = 2.0$ upon the particle size is illustrated in Fig.4.9 where the angular distribution in the scattered intensity is plotted in arbitrary units for different values of the size parameter. The exciting radiation incident upon the particle is assumed to be horizontally polarized. There is a sharp increase in the differential scattering

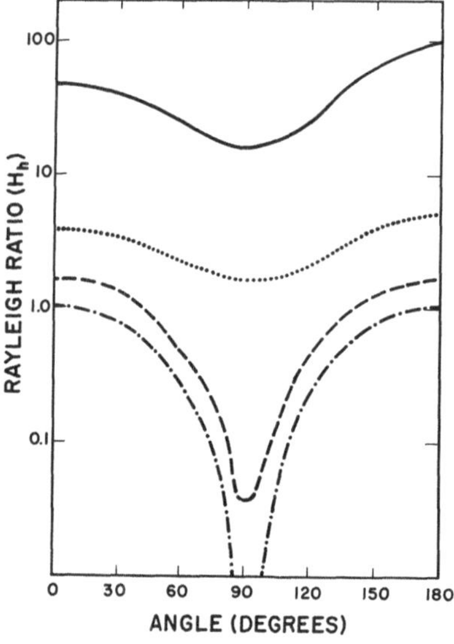

Fig.4.9. Inelastically scattered intensity H_h for spheres uniformly filled with active molecules $\alpha = 5$ (——) $\alpha = 3$ (····), $\alpha = 1$ (----); $\alpha = 0.01$ (-·-·-·-), $m_1 = 2.0$ $\lambda = 1.5\ \lambda_0 = 733$ nm

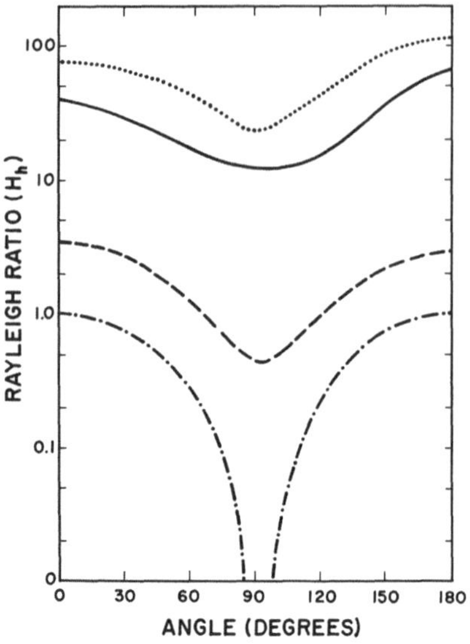

Fig.4.10. Same as Fig.4.9 except $m_1 = 1.5$

and significant changes in the shape of the angular distribution as the particle size parameter α increases from 0.01 to 3 followed by a smaller decrease for $\alpha = 5$.

Figure 4.10 shows the angular distributions for a refractive index of 1.5. The increase in intensity with particle size is less dramatic than shown in Fig.4.9 for $n_1 = 2.0$. However, the trends in the change in shape between very small values of α and $\alpha = 3$-5 are similar to that observed in Fig.4.9 with the exception of $\alpha = 1$. In Fig.4.9 the curves for horizontal incident polarization I_H and I_V have narrowed their differences considerably compared to the corresponding curves in Fig.4.10.

The effect of changing the wavelength of the inelastically scattered wave is illustrated by comparing the curves in Fig.4.9 with the corresponding curves shown in Fig.4.11. The only difference in calculating the curves in the two figures is that in Fig.4.11 the scattered wavelength has been changed from 733 nm to 525 nm. Note that the scattered intensity does not decrease between $\alpha = 3$ and $\alpha = 5$ for $\lambda = 723$ nm as it did for 525 nm.

The angular distributions obtained for inelastic scattering by uniformly filled spheres shown in Figs.4.9-11 are quite different from the corresponding curves for elastic or Lorenz-Mie scattering shown in Fig.4.1. This difference is due primarily to the fact that the dipoles are emitting incoherently while Lorenz-Mie scattering involves coherent scattering by all the molecules comprising the particle. Also there are other essential differences involved in the boundary

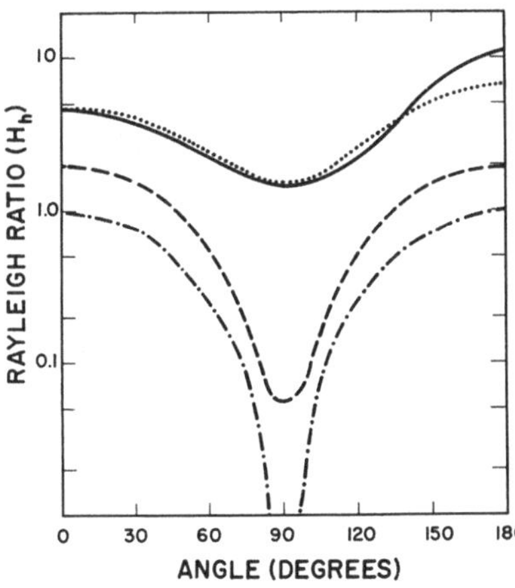

Fig.4.11. Same as Fig.4.9 except λ = 525 nm

value problems in the two cases. Lorenz-Mie scattering involves two fields outside the particle and one (internal) field inside. At the shifted frequency in incoherent inelastic scattering there are two fields inside the dipole field and an induced field needed to satisfy the boundary conditions.

4.3.3 Comparison with Experiments

Experimental measurements of fluorescence by small particles have recently been carried out by KRATOHVIL et al. [4.33] at CLARKSON and LEE et al. [4.34] at Yale. Their data are plotted in Figs.4.12,13, respectively, and are compared with calculations based on the model which has been described above. The curves represent the parallel component of the scattered radiation when the incident radiation is polarized parallel to the scattering plane (H_h) and the perpendicular component of the scattered radiation when the radiation incident on the particle is vertically polarized (V_v). Both theory and experiment deviate from the classic dipole pattern. The deviation for the perpendicular case is relatively small and consists primarily in preferential scattering in the backward direction. Experiments and theory are in good agreement. While the trends are similar for the parallel case the minimum in each experimental curve is not as deep as that in the calculations.

To make a detailed fit of the model with these experiments would require knowing the form that α' takes in (4.18). The model described above assumes that the active molecules were isotropically polarizable and that the particle matrix does not allow the molecules to change their orientation between absorption and emission i.e., α' in (4.18) as a scalar. In general the active molecules have preferred directions of absorption and emission. Moreover, the molecule may change its orientation somewhat before emitting the scattered photon.

In general, therefore, α' in (4.1) is a tensor, the exact form of which depends upon the orientation of the dipole and the dipole absorption and emission characteristics. The molecule's motion, its interaction with its neighbors, and the decay

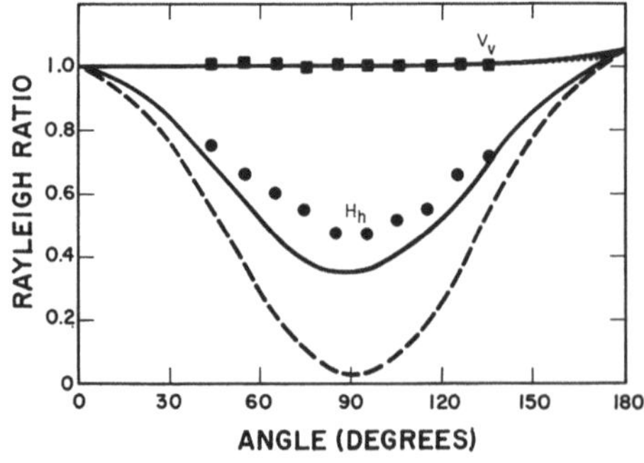

Fig.4.12. Fluorescent scattered intensities H_h and V_v vs angle for polymethylmethacrylatedansylallylamine; modal diameter 304 nm, standard deviation 18 nm, relative refractive index m = 1.13, wavelengths in aqueous medium $\lambda = 277$ nm, $\lambda \geq 376$ nm; based on the data of [4.33], (4.11) and H_h (circles) and V_v (squares) are the polarized components of I_h and I_v which are polarized in the same plane as the incident radiation

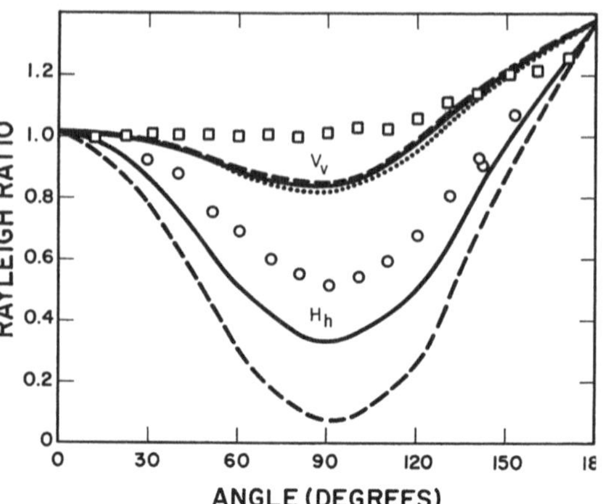

Fig.4.13. Fluorescent scattered intensities H_h and V_v vs scattering angle for 806 nm diameter particle with relative refractive index 1.195 in aqueous medium; $\lambda_0 = 488$ nm $\lambda = 525$ nm; based on the data of [4.34],(4.12)

time all affect the fluorescence. We have recently extended the model to anisotropically polarizable molecules [4.35]. Curves are drawn in Figs.4.12,13 for two examples. The dashed curve represents dipoles that have the same preferred (antenna) direction for dipole transitions in both absorption and emission. The dipoles are assumed to be randomly oriented and this case is called the randomly oriented fixed antenna (ROFA).

The dotted curves in Figs.4.12,13 represent the case of randomly oriented dipoles with uncorrelated emission (ROUE). In this case the absorption and emission dipole directions are completely uncorrelated and the emission from any particular site within the particle is isotropic. The dotted curves in Fig.4.13 show that the radiation pattern for a particle containing a uniform distribution of such sites of isotropic emission may not be isotropic because of the variation of the exciting field within the particle and the refraction of the dipole emission upon emergence from the particle. In Fig.4.13 this leads to enhanced emission in the backward direction.

The experimental data in both experiments lie closer to the ROFA case than ROUE. This is to be expected since the fluorescing molecules in both cases are embedded in solid particles. The ROUE model would be more applicable to fluorescing molecules embedded in liquid drops in which they rotate freely between absorption and emisison.

If the angle between the absorption dipole direction and that of emission is varied to best fit the experimental data, the radiation patterns in Figs.4.12,13 can be fit to within a few percent. However, a definitive test of theory and experiment requires a prior determination of α' in (4.18).

Spheres with Active Molecules Embedded in Surface

In many problems involving remote sensing, the molecules of interest comprise the outer surface of the particle or have been adsorbed onto it. The presence of the particle affects the inelastic scattering to about the same degree as for the uniformly filled particle. Figure 4.14 shows the angular distribution of the H_h components for the same number of dipoles distributed uniformly over the surface of an otherwise uniform dielectric sphere of refractive index 1.5 for different values of α. The minimum at 90° is rapidly filled in. The scattering in the forward and backward directions appear particularly sensitive to particle size.

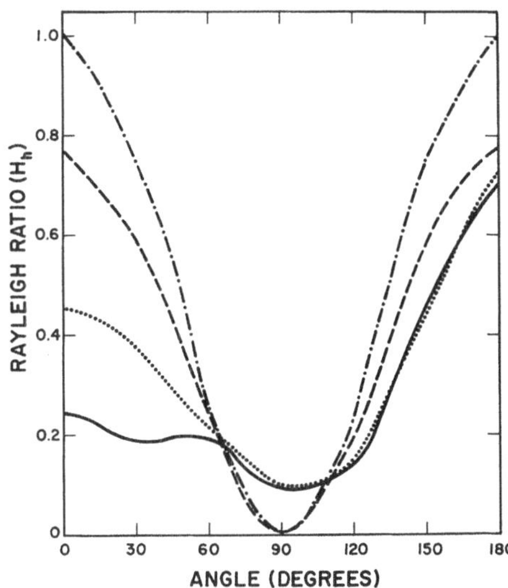

Fig.4.14. Fluorescent scattered intensities I_H vs scattering angle for a sphere coated with active molecules just beneath the outer surface; α = 5 (solid curve), α = 3 (·····), α = 1 (-----), α = 0.01 (-·-·-·-), $m_1 = m_2 = 1.5$

4.4 Coherent Inelastic Scattering

If the active molecules within the particle emit the inelastically scattered light coherently the effects of the particle on the scattering are quite different from those shown above for incoherent scattering. In coherent scattering the inelastically scattered photons maintain a phase relationship with one another. To calculate the scattered field intensity at the location of the observer, the fields must be added prior to obtaining the time-averaged power. One potential field of applications would involve coherent Raman techniques using lasers at two different incident frequencies to produce a coherent output beam. This would be quite complicated to model, but the effects of the particle on the scattering can at least qualitatively be demonstrated by considering the same model as that leading to the expansion coefficients $C_E(\ell,m)$ and $C_M(\ell,m)$ (4.33) which are needed to describe the field corresponding to a particular dipole. The dipoles are assumed in what follows to be distributed uniformly throughout the particle. The scattered radiation field can be obtained analytically for the case of the uniform spherical particle by integrating the field associated with each dipole (4.27,28) over the particle using (4.29-35). The dependence of these fields on the coordinates of the source point $\underline{r}'$ is contained in the coefficients $a_E(\ell,m)$ and $a_M(\ell,m)$. Integration over all $\underline{r}'$ within the particle gives the magnetic field at $\underline{r}$ due to a uniform distribution of scatterers in the particle. Since f_ℓ and g_ℓ in (4.33) are independent of $\underline{r}'$, we have

$$\int \underline{B}_2(\underline{r})d^3r' = [\exp(ikr)/kr] \sum_{\ell,m} (-1)^{\ell+1}\{f_\ell \underline{Y}_{\ell\ell m}(\hat{r}) \int a_E(\ell,m)d^2r' + g_\ell[\hat{r}\times\underline{Y}_{\ell\ell m}(\hat{r})] \int a_M(\ell,m)d^3r'\} \quad . \tag{4.37}$$

We have recently shown [4.36] that

$$\int a_E(\ell,m)d^3r' = \frac{4\pi(\omega/\omega_0)^2\alpha\{j_\ell(k_2a)[k_1aj_\ell(k_1a)]'-n_1^2j_\ell(k_1a)[k_2aj_\ell(k_2a)]'\}}{(n_1^2-1)\{n_1^2j_\ell(k_1a)[k_2ah_\ell^{(1)}(k_2a)]'-h_\ell^{(1)}(k_2a)[k_1aj_\ell(k_1a)]'\}}\,\alpha_E(\ell,m) \tag{4.38}$$

$$\int a_M(\ell,m)d^3r' = \frac{4\pi\alpha(\omega/\omega_0)\alpha_M(\ell,m)}{n_1^2-1}\,\frac{j_\ell(k_1a)[k_2aj_\ell(k_2a)]'-j_\ell(k_2a)[k_1aj_\ell(k_1a)]'}{h_\ell^{(1)}(k_2a)[k_1aj_\ell(k_1a)]'-j_\ell(k_1a)[k_2ah_\ell^{(1)}(k_2a)]'} \tag{4.39}$$

where $\alpha_E(\ell,m)$ and $\alpha_M(\ell,m)$ are the Lorenz-Mie coefficients given in (4.7,8). The total magnetic field at the observer is given by

$$\underline{B}_{TOT}(\underline{r}) = \int \underline{B}_2(\underline{r})d^3r' = [4\pi\alpha\ \exp(ikr)/kr(n^2-1)](-1)^{\ell+1}[f_\ell\beta_E(\ell,m)\underline{Y}_{\ell\ell m}(\hat{r}) + g_\ell\beta_M(\ell,m)\hat{r}\times\underline{Y}_{\ell\ell m}(\hat{r})] \quad . \tag{4.40}$$

This expression is not the same as the Lorenz-Mie amplitude given by (4.7,13,14) as can be seen from the presence of the factors f_1 and g_1. Coherent emission at the shifted frequency is stimulated by the second (the lower) frequency, but in this case there is saturation, i.e., stimulation is independent of field strength.

The coherent inelastic scattering and Lorenz-Mie scattering are compared in Figs. 4.15-17 for three values of the size parameter α. Results are given for the incident radiation polarized both parallel and perpendicular to the scattering plane. There

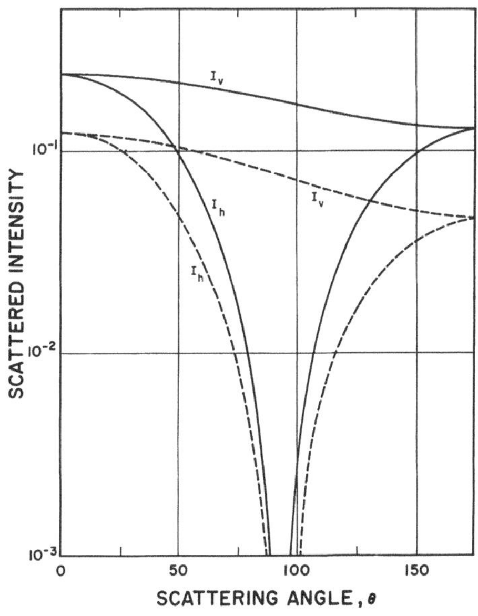

Fig.4.15. Coherent inelastic scattering I_H and the corresponding Lorenz-Mie scattering are plotted vs scattering angle; α = 1 and m_1 = 1.5

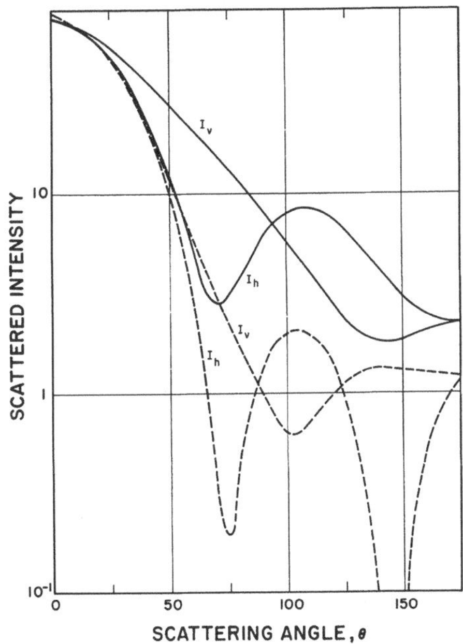

Fig.4.16. Same as Fig.4.15 except α = 3

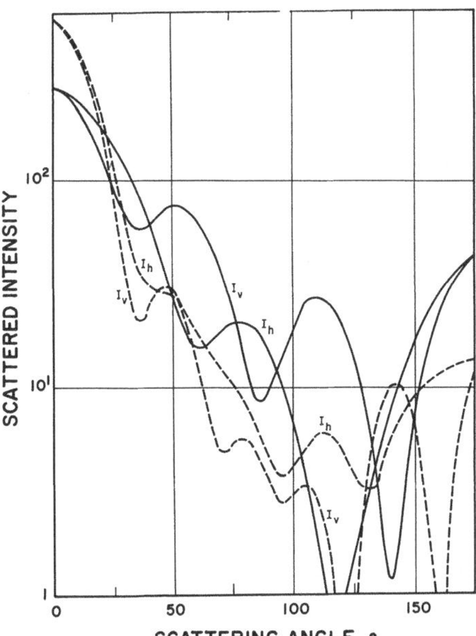

Fig.4.17. Same as Fig.4.15 except α = 5

is qualitative similarity in shape between coherent inelastic scattering and Lorenz-Mie, with the agreement close for particles smaller than α = 1. The scattering patterns for α = 1 are compared in Fig.4.15. The symmetric dipole pattern that holds for very small particles has become enhanced in the forward direction ($\theta = 0^o$) relative to the backward direction ($\theta = 180^o$). This forward to backward asymmetry increases for larger particle sizes, and additional maxima and minima appear in the angular pattern as shown in Figs.4.16,17. The coherent inelastic- and elastic-scattering distributions, although qualitatively similar in shape, exhibit distinct differences.

4.5 Resonance Effects

4.5.1 Absorption

The attenuation of light by a particle includes contributions from elastic scattering and absorption. The absorbed photons induce transitions leading to Raman or fluorescence scattering or contribute to the internal heating of the particle. The extinction cross section is the sum of the elastic-scattering cross sèction and the absorption cross section. The corresponding efficiency factors formed by dividing the scattering cross sections by the geometrical cross section of the particle πa^2 can be written in terms of the standard partial wave amplitudes a_ℓ and b_ℓ [4.2]

$$Q_{ext} = (2/\alpha^2) \sum_{\ell=1}^{\infty} (2\ell + 1)\mathrm{Re}[a_\ell + b_\ell]$$

$$Q_{sca} = 2/\alpha^2 \sum_{\ell=1}^{\infty} (2\ell + 1)(|a_\ell|^2 + |b_\ell|^2)$$

$$Q_{abs} = Q_{ext} - Q_{sca}$$

where Q_{ext}, Q_{sca} and Q_{abs} are the respective efficiency factors for the extinction, elastic scattering, and absorption; and Re denotes the real parts of a_ℓ and b_ℓ.

The amplitudes a_ℓ and b_ℓ are given in VAN DE HULST [4.1] and KERKER [4.2] and can be obtained from (for right circular polarization)

$$a_\ell = -\beta_E(\ell,1)/\alpha_E(\ell,1)$$

$$b_\ell = -\beta_M(\ell,1)/\alpha_M(\ell,1)$$

where $\alpha_M(\ell,1)$, $\alpha_E(\ell,1)$, $\beta_E(\ell,1)$, and $\beta_M(\ell,1)$ are given by (4.3,4,13,14), respectively.

When the refractive index is real, the real part of each scattering amplitude equals the square of its magnitude, the extinction and elastic-scattering efficiency are identical, and the absorption efficiency is zero. The elastic-scattering efficiency Q_{sca} increases with increasing α to a maximum of 3-6 and then undergoes damped oscillations about the limiting value $Q_{sca} = 2$. Superimposed upon this main oscillation is a so-called ripple structure which becomes increasingly irregular at higher refractive indices. The ripple structure in the Q_{sca} dependence on α is the result of resonances in the partial wave amplitudes a_ℓ and b_ℓ [or $\beta_E(\ell,1)$ and $\beta_M(\ell,1)$] [4.2]. For small values of ℓ the contributions from the various amplitudes overlap but with increasing ℓ they became sharper and more distinct, particularly for the magnetic contributions. In general the ripples are too large to ignore, particularly for large values of the refractive index. According to CHYLEK [4.37], the first few sharp resonances in each partial wave are connected with surface waves.

The ripples in Q_{ext} are in general less pronounced for complex refractive indices. CHYLEK et al. [4.38] have shown that as the imaginary component of the refractive index increases the height of the resonance peaks are reduced and their widths increased. The first-order resonances in a_ℓ and b_ℓ are most sensitive to increases in the imaginary component of the refractive index and practically disappear for even small values. The contributions from higher-order resonances in a_ℓ and b_ℓ remain to give Q_{abs} a resonance-like structure for particles comprised of weakly absorbing dielectrics. BENNETT and ROSASCO [4.39] having studied these resonances in the partial wave amplitudes in some detail point out that resonances in the partial wave amplitudes are physically resonances in the energy density inside the spherical particle and should, therefore, correspond to resonances in fluorescence and Raman scattering.

Examination of the partial wave amplitudes $C_E(\ell,m)$ and $C_M(\ell,m)$ for fluorescence and Raman scattering in the model of CHEW et al. [4.10] given by (4.33-35) do, in fact, show the same factor in the denominator as a_ℓ and b_ℓ. Therefore, resonances in a_ℓ and b_ℓ that result from the denominators approaching zero are likely to correspond to resonances in $C_E(\ell,m)$ and $C_M(\ell,m)$.

FUCHS and KLIEWER [4.40] point out that these peaks in the Mie scattering efficiencies can be identified with the virtual modes of the sphere.

Recently, ASHKIN and DZIEDZIC [4.41] have reported resonances in the radiation pressure required to levitate dielectric spheres against the force of gravity by means of a laser beam. They used a technique based on optical levitation which they call "force spectroscopy". Individual micron-sized particles are trapped in a laser beam with a Gaussian intensity profile and levitated by varying the laser power. The light power from a tunable dye laser needed to hold the sphere fixed in space determines the radiation pressure. In a series of measurements of the radiation pressure needed to levitate particles in which the wavelength of the incident light and the size of the particle were varied ASHKIN and DZIEDZIC [4.41] report resolutions of $\Delta\lambda/\lambda = \Delta\alpha/\alpha = 4 \times 10^{-5}$. CHYLEK et al. [4.42] have analyzed this radiation-pressure data and have shown that the positions of the sharp peaks observed by ASHKIN and DZIEDZIC [4.41] are in excellent agreement with the partial wave resonances described above. The technique of optical levitation developed by ASHKIN [4.43] has considerable potential in that it provides a technique for positioning particles for light-scattering measurements on individual particles in liquids, as well as in air. Force spectroscopy may provide a technique for size determination with considerably better resolution than far-field scattering techniques and provide an exacting test of the Lorenz-Mie theory [4.41].

4.5.2 Active Particles

Some interesting resonance effects on scattering by "active" particles have been pointed out recently by ALEXOPOULOS and UZUNOGLU [4.44] and amplified by KERKER [4.45]. Normally, scatterers have indices of refraction with small negative imaginary parts which are responsible for absorption. However, a scatterer containing molecules which have been "pumped" by a laser beam to attain inverted populations of energy levels may act as an amplifier when excited by light of the appropriate frequency. In this case the scatterer may be ascribed a refractive index with a positive imaginary part [time convention $\exp(i\omega t)$] which would then amplify rather than absorb the incident light. Such systems have many interesting properties [4.44,45]. For example, as is well known, the moduli of the Mie coefficients in the normal case cannot exceed unity as a consequence of unitarity. When the refractive index has a negative imaginary part, however, the unitarity bound does not apply, and the moduli of the Mie coefficients may attain values far in excess of unity, and arbitrarily sharp resonances are not excluded (as they are in the normal case). For numerical studies of these and other effects, see KERKER [4.46].

4.5.3 Evanescent Waves

The study of scattering of evanescent electromagnetic waves is attracting increasing interest. Experimental studies on the angular distribution of fluorescence from thick liquid dye layers excited by evanescent waves have been made by LEE et al. [4.47]. In these experiments, layers of micron-sized spherical particles containing fluorescent scatterers are immersed in a liquid and placed on the flat face of a prism through which an evanescent wave is generated by total internal refraction. The results were found to be in good agreement with Fresnel theory.

Calculations of elastic scattering of evanescent waves by dielectric spheres sufficiently far from the media boundary so that they do not perturb the plane boundary conditions have been made by CHEW et al. [4.48]. These calculations show significant departure from Mie scattering even when the damping is very small. There are also interesting polarization effects not found in the scattering of plane waves. For example, if the scattering plane is parallel to the media boundary, the intensity of the spin-flip component of the scattered beam (which is zero for Mie scattering) is appreciable in many cases even when the damping is very small.

4.6 Summary and Conclusions

Fluorescent and Raman scattering combined with Lorenz-Mie scattering have the potential for nondestructive analysis of small particles to sizes well into the submicron range. Using Raman and fluorescent scattering the experimenter can isolate molecules of a specific type within the particle. The optical and geometrical properties of the particle and the location of the active molecules within the particle will strongly affect the intensity, angular distribution, and even the emission spectra observed at a given scattering angle far from the particle.

These effects have been illustrated by calculations for simple idealized cases. There appear to be quite pronounced effects due to the particle and they must be taken into account in any application of inelastic scattering for quantitative determinations of the amount of a specific molecule in an aerosol. Model calculations of inelastic scattering by isotropically polarizable electric dipoles uniformly distributed within an otherwise nonabsorbing dielectric sphere exhibit a variation of more than two orders of magnitude in the inelastically scattered intensity per active molecule as the particle size is varied.

The model of inelastic scattering described in this chapter is based upon the assumption that the active molecule can be represented by a classical oscillating electric dipole whose strength is determined by the strength of the local field at the exciting frequency given by the Lorenz-Mie theory. These assumptions should be reasonable for many molecules embedded in weakly absorbing particles. Recent experiments on fluorescence confirm the qualitative features predicted by the model. A rigorous experimental test has yet to be performed. Such a test would require knowledge of the active molecules' scattering characteristics when embedded in bulk material identical to that comprising the particle. However, these scattering characteristics can alternatively be determined by scattering measurements performed on particles of the same composition that have diameters much smaller than the wavelength of the incident light.

Acknowledgment

We gratefully acknowledge the assistance of Dr. Derry Cooke, Dr. Steven Druger, and Mr. Michael Sculley with the numerical calculations. This work was supported in part by DOE Contract No. EE-77-S-02-4361 and NSF Grant CHE 77-13102.

References

4.1 H.C. Van de Hulst: *Light Scattering by Small Particles* (Wiley, New York 1964)
4.2 M. Kerker: *The Scattering of Light and Other Electromagnetic Radiation* (Academic Press, New York 1965)
4.3 M. Born, E. Wolf: *Principles of Optics* (Pergamon Press, New York 1959)

4.4 D. Miller, G. Benedek: *Intraocular Light Scattering* (Charles C. Thomas, Springfield, Illinois 1973)
4.5 See for example the proceedings of the Fourth and Fifth Conference on Automatic Cytology, J. Histochem. Cytochem. *24*, 1-414 (1976); *25*, 479 (1977)
4.6 R.T. Collis, M.G. Ligda: Nature *203*, 508 (1964)
4.7 M. Delhaye, P. Dhamelincourt: J. Raman Spectrosc. *3*, 33 (1975)
4.8 E.S. Etz, G.J. Rosasco: In *Environmental Pollutants*, ed. by T.Y. Toribara, J.R. Coleman, B.E. Dahneke, I. Feldman (Plenum Press, New York 1978) pp.413-456
4.9 G.J. Rosasco, E.S. Etz, W.A. Cassatt: Appl. Spectrosc. *29*, 396 (1975)
4.10 H. Chew, P.J. McNulty, M. Kerker: Phys. Rev. A*13*, 396 (1976)
4.11 H. Chew, M. Kerker, P.J. McNulty: J. Opt. Soc. Am. *66*, 440 (1976)
4.12 J.D. Jackson: *Classical Electrodynamics*, 2nd ed. (Wiley, New York 1975)
4.13 A.R. Edmonds: *Angular Momentum in Quantum Mechanics* (Princeton University Press, Princeton, New Jersey 1968)
4.14 A.L. Aden, M. Kerker: J. Appl. Phys. *22*, 1242 (1951)
4.15 M. Kerker: J. Opt. Soc. Am. *65*, 375 (1975)
4.16 H. Chew, M. Kerker: J. Opt. Soc. Am. *66*, 445 (1976)
4.17 P. Morse, H. Freshbach: *Methods of Theoretical Physics*, Vol.2 (McGraw-Hill, New York 1953)
4.18 V.A. Erma: Phys. Rev. *179*, 1238 (1969)
4.19 R. Mittra (ed.): *Computer Techniques for Electromagnetics* (Pergamon Press, Oxford 1973)
4.20 P.L. Uslenghi (ed.): *Electromagnetic Scattering* (Academic Press, New York 1978)
4.21 P.C. Waterman: Phys. Rev. D:*3*, 825 (1971)
4.22 S. Ström: Phys. Rev. D:*10*, 2685 (1974)
4.23 P. Barber, C. Yeh: Applied Optics *14*, 2864 (1975)
4.24 P.C. Waterman: Matrix Methods in Potential Theory and Electromagnetic Scattering (to be published)
4.25 D. Saxon: UCLA Lecture Notes (unpublished)
4.26 R.G. Newton: *Scattering Theory of Waves and Particles* (McGraw-Hill, New York 1965)
4.27 S. Strom: Am. J. Phys. *43*, 1060 (1975)
4.28 M. Kerker, D. Cooke, H. Chew, P.J. McNulty: J. Opt. Soc. Am. *68*, 592 (1978)
4.29 J.J. Bowman, T.B.A. Senior, P.L.E. Uslanghi: *Electromagnetic and Acoustic Scattering by Simple Shapes* (North-Holland, Amsterdam 1969)
4.30 S. Asano, G. Yamamoto: Appl. Opt. *14*, 29 (1975)
4.31 C. Flammer: *Spheroidal Wave Functions* (Stanford Univ. Press, Stanford 1957)
4.32 M. Kerker, P.J. McNulty, M. Sculley, H. Chew, D.D. Cooke: J. Opt. Soc. Am. , 1676 (1978)
4.33 J.P. Kratohvil, M.P. Lee, M. Kerker: Appl. Opt. *17*, 1978 (1978)
4.34 E.H. Lee, R.E. Benner, J.B. Fenn, R.K. Chang: Appl. Opt. *17*, 1980 (1978)
4.35 P.J. McNulty, S.D. Druger, M. Kerker, H.W. Chew: Appl. Opt. *18* (1979)
4.36 H. Chew, M. Sculley, M. Kerker, P.J. McNulty, D.D. Cooke: J. Opt. Soc. Am. *68*, 1686 (1978)
4.37 P. Chylek: J. Opt. Soc. Am. *66*, 285 (1976)
4.38 P. Chylek, G.W. Grams, R.G. Pinnick: Science *193*, 480 (1976)
4.39 H.S. Bennett, G.J. Rosasco: Appl. Opt. *17*, 491 (1978)
4.40 R. Fuchs, K.L. Kliewer: J. Opt. Sco. Am. *58*, 319 (1968)
4.41 A. Ashkin, J.M. Dziedzic: Phys. Rev. Lett. *28*, 1351 (1977)
4.42 P. Chylek, J.T. Kiehl, M.K.W. Ko: Phys. Rev. A:*18*, 2229 (1978); Appl. Opt. *17*, 3019 (1978)
4.43 A. Ashkin: Phys. Rev. Lett. *24*, 156 (1970)
4.44 N.G. Alexopoulos, N.K. Uzunoglie: Appl. Opt. *17*, 235 (1978)
4.45 M. Kerker: Appl. Opt. *17*, 3337 (1978)
4.46 M. Kerker: Appl. Opt. *18*, 1180 (1979)
4.47 E.H. Lee, R.E. Benner, J.B. Fenn, R.K. Chang: Appl. Opt. *18*, 862 (1979)
4.48 H. Chew, D.-S. Wang, M. Kerker: Appl. Opt. *18*, 2679 (1979)

5. Survey of Aerosol Interaction Forces

W. H. Marlow

With 1 Figure

The forces of interaction (i.e., prior to contact) which a single, gas-borne particle can be subject to are treated from the perspective of its chemical and physical structure. To provide the requisite perspective for understanding the importance of these compositionally dependent factors, the role of the gas is discussed. Classical electrostatic and multipolar forces and the thermodynamic setting for any interaction involving a particle are described briefly. Principle emphasis in the chapter is given to the van der Waals forces. The modern (Lifshitz) theory is introduced and its relation to the classical Hamaker theory is described. A qualitative discussion of the computational approaches commonly used and experimental evidence for the theory are given. Inclusion of the chemical and physical factors necessary for treatment of cases that arise in actual application of the general theory is discussed.

5.1 The Chemical Physics of Aerosol Particle Interactions

In the conventional treatment of the dynamics of aerosol processes, it is customary to assume that all collisions result in adhesion (or coagulation) and that high electrostatic particle charge is the only effective retardant or accelerator of these processes. These assumptions are based upon experimental observations of particle coagulation, calculations of the strengths of particle interaction forces of various kinds, and knowledge of particle diffusion, or Brownian motion. As far as it goes, this experimental and theoretical evidence remains valid and will not be reviewed here because of its coverage elsewhere [5.1-3].

During the last decade or so, new knowledge has emerged from research both within and outside aerosol investigations that refines the conventional picture's domains of applicability and extends the understanding of aerosol phenomena. These considerations have important implications for very practical questions some of which were discussed in Chap.1. Likewise, model calculations on transport to transition and free-molecular regime particles clarify the importance of the details of particle interaction forces and particle transport and may have implications for coagulation, condensation, and sorption processes of aerosols.

These recent advances in the understanding of particle interaction forces are due to improved definition of the pertinent variables. Therefore, this review will begin with an inventory and discussion of the variables affecting aerosol particle interaction forces.

5.1.1 Scope of the Variables Determining Aerosol Interaction Forces

The effective force a particle experiences is due to factors both of its constitution and of its environment.

Particle chemical composition is determined by the specific system being examined. Therefore, it is important to understand when composition may or may not be overlooked in deference to information on particle behavior arising from data collected on different systems. Overall composition is not of itself adequate because particle structure, morphology, and size play important roles. These questions will all be discussed.

Environmental influences upon the effective force experienced by a particle include the suspending gas, nearby condensed phases, and external fields. The suspending gas is generally the dominant factor in aerosol mechanics and is, therefore, accorded prime status in most monographs [5.1,3]. Its influence is exerted by anisotropy such as in the case of a flowing gas and by inhomogeneity such as the local effects of acoustic waves. When spacial gradients in composition exist in the gas, this too will have an effect and is called Stephan Flow [5.3]. If the gas species for which the gradient is established reacts inelastically in any manner with the particle, a related force upon the particle arises [5.4]. As intensive state variables of the gas, temperature, pressure, and number density have an important local effect upon all particle processes since they determine gas and particle mean free paths, temperature-dependent interactions, relaxation times, etc. [5.5]. The role of the gas will be addressed in this chapter primarily through the Knudsen number, $Kn = l/a$ (l = gas molecular mean free path, a = particle radius), and gas molecular mean free path as indications of the range of effectiveness of the particle interaction forces. Nearby condensed phases such as other particles and surfaces are essential in descriptions of particle forces, both because of their discrete effects such as gas-phase anisotropy and particle interactions and their collective effects which can modify the interaction. The final category of environmental influences upon particle interaction forces is that due to macroscopic electric and magnetic fields. Static electric fields can couple to essentially all particles, either charged or neutral. In the former case, the force is given by the coupling of the charge to the local electric field while for the neutral particle, multipolar coupling or induced-moment coupling occurs and has been studied extensively. Another variety of coupling is also possible if the particle is irradiated either by a laser field at the frequency of an atomic or molecular resonance of the particle (as in Raman or fluorescence spectroscopy described in Chap.4) or by any other field capable of coupling to the particle's electrons.

From the foregoing discussion it can be seen that the forces acting upon aerosol particles divide into two categories: gas dynamics and chemical physics. In all cases, the former moderates the later to determine the spacial extent of its importance. This chapter will be devoted principally to the chemical physics of aerosol interaction forces (see Chap.2 for discussion of the gas dynamic aspects).

5.1.2 Classification of the Chemical Physical Dependence

The Thermodynamics of Aerosol Interactions

When the forces between a particle and another condensed species are treated, thermodynamics and statistical mechanics of the aerosol (particles *plus* gas) enter through temperature dependence of the interaction forces. However, actual aerosol particle interaction forces may be altered in a fundamental way if one or both of the particles or surfaces absorb molecules from the suspending gas. ASH et al. [5.4] considered nonionic systems in which the relative velocity of the particle and surface or other particle is "sufficiently small, in relation to the rates of absorption and desorption, that absorption equilibrium is maintained as the particles move together, collide and then either adhere or separate." They, therefore, assume constant temperature for the entire aerosol system implying at least several nonabsorbing gas molecular collisions with the sorbent species between each sorbate interaction; that is to say, the sorbate must be a minority ($\lesssim$ 10 percent) species in the gas. By use of conventional equilibrium thermodynamics they derive the expression for the excess force (beyond van der Waals and electrostatic) between two bodies due to sorption as

$$F - F^0 = \int_{-\infty}^{\mu} \left(\frac{\partial n^\sigma}{\partial D}\right)_{T,\mathrm{shape},\mathrm{size}} d\mu \quad . \tag{5.1}$$

Here, μ is the chemical potential of the vapor over the (similar) surfaces of the sorbents, n^σ is the total amount absorbed, D is the distance of separation. In this case, $d\mu$ is taken to be given in terms of the vapor fugacity relative to the fugacity at a standard pressure. The interpretation given this result is that if in moving the two bodies together an increasing number of molecules become absorbed (e.g., the species evaporated from one is absorbed by the other body), then attraction occurs since the vapor-absorbent system reaches a lower overall potential. Conversely, if moving the bodies together lowers the density of absorbate molecules available between the two bodies, then repulsion can occur. Since the intermolecular interaction force term F^0 may be substantially affected by the sorbed layer, thermodynamic arguments such as these can only serve to delineate the behavior of these systems under special conditions.

Interaction Forces

Forces among molecules beyond the orbital overlap zone, and more generally among clusters and particles in the present case, are customarily classified [5.6] as Coulomb, Keesom, Debye-Falkenhagen, and van der Waals. In the Coulomb case, one or both species carries an excess charge which dominates the interaction, either via Coulomb's law or by the charge-induced image force to be discussed briefly. Keesom forces are the multipole-multipole interaction forces between neutral bodies with distributed fixed charges. Though undoubtedly important under some situations such as very close encounters where rotational Brownian motion (which would tend to cancel the interaction) does not enter, their overall importance is difficult to assess without details of the multipoles, information generally unavailable. In any case, once these multipoles are known, the interaction energies can be computed from standard formulas of electrostatics. Multipole-induced multipolar interactions are known as Debye (or Falkenhagen) forces and like Keesom forces are important for bodies possessing permanent moments when interacting with polarizable bodies.

The more important of the intermolecular forces is the van der Waals force, a quantum-mechanical interaction arising in large measure from the zero-point fluctuations of the interacting media's electrons. Since the interaction force is the result of fluctuation-induced fluctuations, it can be compared with the other induction forces. In essentially all cases [5.7] the attraction due to a monopolar charge induced image force dominates van der Waals forces. Multipole-moment induced forces can not generally be expected to play an important role in the aerosol domain. For a neutral particle to have moments, it must have separated, distributed charges which restricts its possible composition to electrical insulators. A gas-phase bipolar ion will tend to attach to the particle at or near where the isolated charges resides, thereby diminishing the possible multipole moment.

Chemical Dependency of Aerosol Interaction Forces

Besides the importance of the specific molecular species interacting with the particle, which properly falls under the domain of heterogeneous nucleation, particle chemistry is of fundamental importance in the domain of interaction forces.

In electrostatic interactions, the particle image force plays a central role in particle charging, the acquisition of cluster ions by a dielectric particle [5.8,9]. For all condensed media, the static dielectric constant is intimately related to the molecular polarizability of its constituents as collective and structural effects. This situation can be made more complicated if a semiconducting particle exhibiting the "Zener effect" allowing it to change polarizability upon approach of a charge is considered. In any case, the origin of the charge-particle interaction rests in the particulate chemical species via its physical properties.

The interactions of uncharged species have been touched upon above. Since the van der Waals forces dominate these interactions, they will be discussed at length in the final section of this chapter. It suffices here to say that these forces arise from the frequency-dependent electric and magnetic susceptibilities of the interacting species, and it is precisely these susceptibilities which are responsible for the spectral properties of the molecules comprising the particle. Thus, molecular (or chemical) specificity of particle interaction forces is of central importance. From another standpoint, this can be understood by considering an individual molecule as the limiting case of a particle. Then for the intermolecular van der Waals force to be consistent as two such particles (molecules) approach to the point of orbital overlap, their interaction must reduce to the relevant chemical interaction force which is fundamentally dependent upon chemical specificity.

Particle Geometry in Aerosol Interaction Forces

In addition to the details of chemical composition which characterize the particle interaction force, considerations of shape, size, and gas density also must be included. For example, the polarizability of a neutral pair of adhering spheres is greater along its axis of rotational symmetry than in any other direction. Therefore, its interaction energy is greatest along that axis with the result that a third sphere will preferentially attach itself to one end to form a "chain agglomerate" rather than in the middle to form a triangle (see Sect.5.4.6 and references therein). Potentially, this has a considerable effect upon all varieties of interactions since it determines the orientation of particles which adhere to surfaces and may, therefore, influence subsequent chemical and physical reactions through exposure, or concealment, of active surfaces or sites.

Gas density enters the question of interaction forces indirectly by modifying the particle's trajectory in the potential field of the matter it interacts with. There are highly approximative means of dealing with this for spheres where the Knudsen number regime is well defined. In the case of other shapes such as rods which may span several ranges of Kn, the effect can become impossibly complicated when orientation dependency of the forces is included.

The strength, and attractive or repulsive character, of the interaction force is highly dependent upon distance from the particle. Therefore, the dimensions of the particle in relation to the gas molecular mean free path plays an important role in determining the effective force upon a particle.

Size Effects and Particle Interactions

As the condition of matter between isolated atoms or molecules and matter-in-bulk, clusters and fine particles display many physical properties which differ considerably from those of the bulk states of the same material [5.10]. This field of current research is reviewed in [5.11]. Of particular interest in the current considerations is the departure of the dielectric constant from its bulk-state value due to the increased fraction of molecules on or next to the surface 5.12 . Besides their importance in van der Waals interaction forces already described, these surface effects may also be responsible for aspects of the well-known greatly increased catalytic activity of clusters and very fine particles compared with bulk matter. Because the detailed chemical physics of aerosols must ultimately devolve to the physical properties of the individual particles, the interactions of particles with other particles, gases, and surfaces are all dependent upon increased understanding of microparticle microphysics.

Composite Particles and Their Interactions

Composition alone is insufficient to characterize a particle's interactions since the material of which it is composed can be arranged in different configurations within it. In many cases, this arrangement is known from the particle's formation processes or its environmental conditions so that its interaction forces can be calculated. The electric and magnetic susceptibilities of layered structures are of particular importance in the van der Waals interaction between particles or with a substrate onto which it might be expected to be deposited. Since the particles under consideration are very small, the layering may be expected to show effects of finite thickness whose physical properties vary with film thickness and substrate composition, as is known in other contexts.

5.1.3 Material to Be Covered

This survey consists of three parts in which an attempt is made to describe the unique aspects of interaction forces experienced by particles of Kn $\gtrsim$ 1. As such, no unified theory presently exists so results of a detailed model calculation are presented in the next section for the development of an intuitive picture of how kinetic theory exerts its influence on aerosol interactions.

The second section gives a brief recounting of some recent papers on electrical interactions appropriate for inclusion in aerosol considerations.

Van der Waals forces are generally the most neglected ones in aerosol studies, a peculiar situation in light of the pragmatic orientation of most aerosol considerations which perforce encounter particles of complex structure. The last section of this chapter surveys this subject pointing to results that should be of especial value in aerosol work.

This survey does not consider the interactions of aerosol particles once they are in "contact." For this reason, the electrostatic effects of mobile charge species in uncharged particles are not treated. However, they are of central importance in considerations of the chemical and structural evolution of the contacting particles whenever they contribute to double layer formation. In that case, they play a major role in determining the stability of the coagulated particles via considerations of the balance between forces arising from the double layer and van der Waals interactions. Under those circumstances where the minimum of the potential-energy curve with respect to separation is at contact with no interpenetration of the surfaces, the agglomerated particles might fall apart because of molecular collisions.

5.2 Kinetic Theory and Aerosol Interaction Forces

5.2.1 The Physical Context for Particle Interactions in the Subcontinuum Zone

Transport processes involving aerosol particles are frequently classified according to the particle Knudsen number [Ref.5.3, Chap.1]. When a particle is much larger than a gas molecular mean free path ($Kn \ll 1$), the particle is said to be in the continuum regime since the usual hydrodynamic equations for fluid flow pertain. To the extent that it need be addressed directly, this regime is not discussed in this chapter. The principal focus here is upon particles for which $Kn \gtrsim 1$.

Particle Near a Plane

In the presence of a macroscopic extended surface, the Knudsen number picture can illuminate transport processes especially well in the zone extending from the surface to a small number of mean free paths away from it. Then a particle larger than that zone does not experience any substantial discontinuity in physical properties of the gas as it approaches the surface and its transport there is accounted for by use of a phenomenological slip boundary condition [5.3]. If $10 \gtrsim Kn \gtrsim 1$, the particle is in the *transition regime* of kinetic theory and its transport within the boundary zone of the surface must include the discontinuity of gas properties because of the geometrical asymmetry of gas molecular interactions with the particle. The $Kn > 10$ case is termed the *free-molecular* regime of kinetic theory. It corresponds to those conditions in which the particle size is much smaller than the boundary zone so that the particle can be treated independently of the gas. Evidently, then, if the interaction force's range is less than the thickness of the boundary zone, only a transition regime or smaller particle is significantly perturbed from its gas-dynamic motions by the interaction force. Since such short-range forces become significant only at very close approach as the precursor to adhesion (or as a disjoining force), they are implicit to the slip boundary condition.

Adjacent Particles

The interaction of two particles is subject to a similar description as a single particle at a boundary. The additional complication is due to the partial, rather than complete, shielding of one particle from the gas because of the presence of the other particle. In this picture, a continuum regime particle is similar to a surface because a smaller particle or a molecule interacts with it as if it were a surface. Transport of comparably sized continuum regime particles in each other's vicinity generally is affected only by hydrodynamic interaction and surface forces whose ranges are much smaller than a typical particle dimension unless they are very highly charged. This effect is due to the considerable inertia of these particles and the weakness of most interaction forces as discussed above.

Standard Approach to Coagulation

The classic treatment of aerosol interaction and coagulation is given by FUCHS [5.1] who calculated the coagulation coefficient as

$$K(r_1,r_2) = 8\pi \frac{r_1+r_2}{2} \frac{D_1+D_2}{2} \beta \quad . \tag{5.2}$$

Here, r_1 and r_2 are the particle radii, D_1 and D_2 are their diffusion coefficients, and β is a factor which estimates the effects of both the discontinuity of the gas for transition regime and smaller particles and the collisional sticking efficiency. In the derivation of $K(r_1,r_2)$, the assumption is made that a particle coming within a certain distance, related to its mean free path, of another particle enters a "jump" region (the "dividing sphere") and collides with the target particle. He found that K was relatively insensitive to the sticking coefficient for slip flow and continuum regime particles. Following Smoluchowski's early work, FUCHS also found proportionality between the coagulation constant and the sticking coefficient for free-molecular and near-free-molecular regime particles. Since the dividing sphere of FUCHS may fall 100 nm outside the target particle, complicated interaction potentials whose deviations from monotonicity occur within that distance can be used to estimate effective sticking coefficients.

For longer-range interaction forces, FUCHS uses classical diffusion theory to derive formulas for the flux of particles to a target particle. These formulas were discussed for several monotonic particle interaction potentials and he found that in most cases the role of the interaction force was not significant.

There are several interrelated questions with these approaches, some of which are not generally appreciated and will be discussed in the context of a model calculation below. The question of the coagulation coefficient was also discussed in Chap.2. The most critical shortcoming of this approach is the assumption that all particles at the surface of the dividing sphere for the target particle will collide with it. Exact particle trajectory calculations must be performed if conservation of momentum and energy are not to be violated. This is very critical where short range forces are involved. For long-range repulsive interactions, the two particles can collide only if they have both a small impact parameter and sufficient energy to approach within the distance where van der Waals, adhesion, bonding, or other attractive surface forces can come into effect. The second problem lies in the use of the classical diffusion equation to estimate the importance of specific interaction forces. While this is appropriate for the particles if they are sufficiently separated that their gas-phase environment is essentially isotropic, approach closer than a gas molecular mean free path vitiates the utility of this formula (Chap.2).

In the transition and free-molecular regimes, the difficulty in describing effective aerosol interaction forces lies ultimately in the intractability of the Boltzmann (or other appropriate) kinetic equation to exact solution. In the case of two transition-regime spheres, with absolutely no interaction potential, an effective attractive force arises because the zone of isotropic gas molecular collisions for each particle is truncated by the presence of the other particle. It is this effective interaction force which the dividing-sphere method approximates by assuming complete absorption for distances less than some distance defined for each pair of spheres regardless of their composition.

The difficulties that arise in the description of transition-regime and smaller-aerosol-particle physics may be addressed by the appeal to both limiting cases and model calculations.

By allowing one of the spheres to shrink to a point and assuming its motion to be similar to a gas molecule's, the coagulation problem is rendered slightly more tractable. In the case of only surface interaction, the problem reduces to that of condensation or sorption. Conversely, if longer range potentials, which are not necessarily monotonic such as the electrostatic interaction with a polarizable aerosol particle are considered, the system can be taken to model one of two important aspects of realistic two-particle interactions. This aspect is the role of particle size relative to gas density in the determination of an interactant's collision or repulsion. The other aspect which cannot yet be addressed, is that of attraction due to mutual shielding discussed above.

5.2.2 Model Calculation

Solution of the appropriate kinetic equations for transition-regime particles has not yet been done for cases involving realistic aerosol interactions. Consequently, recourse needs to be taken to model kinetic equations and their solutions to gain insight into the simultaneous roles of both kinetic theory and long- and short-range forces upon transition-regime transport phenomena. The strength of this approach is the clear delineation of relevant parameters while its weaknesses may be numerous, including uncertainty of the extent and possibly even the domain of validity.

One model equation whose solution has proved useful in many cases is the constant collision frequency, relaxation model of the Boltzmann equation which, for spherically symmetric problems, in spherical coordinates is

$$v_r \frac{\partial f}{\partial r} + \frac{v_t^2}{r} + F_r \frac{\partial f}{\partial v_r} - v_r v_t \frac{\partial f}{\partial v_t} = \lambda(f^0 - f) \quad . \tag{5.3}$$

Here, v_r and v_t are, respectively, the nondimensionalized radial and transverse components of velocity, f is the distribution function for the species being described, and f^0 is the local Maxwellian distribution. Charging of spherical aerosol particles of radius a by gaseous ions is modeled by solving the equation with the surface of the sphere placed at r = a and the force F_r appropriately chosen. The distribution function must then be integrated over r = a to give the ion flux.

With

$$\lambda = \sqrt{\pi}(a/\ell_i) \frac{M_i+M_j}{M_j} \tag{5.4}$$

as the collision frequency, ℓ_i the ionic mean free path, M_i the ionic mass, and M_j the mass of a neutral molecule of the gas, (5.3) approaches the free-molecular equation as $Kn^{-1} = (a/\ell_i) \to 0$ since $f \to f^0$. Thus, as a correction to free-molecular transport, this equation may be expected to indicate overall functional and physical dependences. MARLOW and BROCK [5.9] have solved this model equation by an iterative expansion to first order in a/ℓ_i

$$f = f_0 + (a/\ell_i)f_1 \tag{5.5}$$

using the unipolar ion-particle interaction force

$$F_r = \frac{e_1 e_2}{r} - \frac{e_1^2}{2} \frac{\kappa-1}{\kappa+1} \frac{2r^2-1}{r^3(r^2-1)^2} \tag{5.6}$$

which is nondimensionalized according to

$$e_1 = \frac{\varepsilon_i}{(akT)^{\frac{1}{2}}} \quad , \quad e_2 = \frac{\varepsilon_a}{(akT)^{\frac{1}{2}}}$$
$$e_1' = e_1 \frac{\kappa-1}{\kappa+1} \quad , \quad r = \rho/a \tag{5.7}$$

where ε_i and ε_a are respectively the ion and aerosol charges and ρ is the distance from the center of the sphere. F_r is an approximation to the image force on an ion which polarizes a uniform sphere of static dielectric constant κ having a charge ε_a at its center. For the present purposes, it is the shape of this force which is of interest. The radial force F_r is clearly divided into a long-range repulsive Coulombic term and a short-range attractive image term whenever ε_i and ε_a are the same sign.

Computation of the ionic flux to a particle, which is assumed fixed in space [5.13], may be done by the method of Knudsen iteration [Ref.5.14, Chap.II and Appendix A, Case I]. In this procedure, (5.5) is substituted into (5.3) to derive a sequence of equations. The first is the collisionless Boltzmann equation and the second is an equation for f_1 in terms of the solution of this collisionless Boltzmann equation and the ionic distribution in the absence of all particles and fields. Solution of this first iterated equation is calculated by integration over its constant characteristics (total energy and angular momentum). In practice it is useful also to simultaneously integrate over the particle's surface to determine the ionic flux. A component of this computation is the determination of the collisionless (i.e., free-molecular) ionic density as a function of the *sign* of the radial velocity, a procedure which includes the finite size of the aerosol particle. The dividing sphere method discussed above does not differentiate between the two signs of the radial velocity by its use of the "jump distance". Recognition of this difference is equivalent to the partial shielding of the point under examination from the gas by the particle when molecular collisions are included via the right hand side of (5.5). A second aspect of this computation that need be noted is the divergence of the model image potential in (5.6) at the surface of the sphere r = 1. This is handled by a method dependent upon the assumption that the ion distribution assumes its asymptotic value at a distance beyond a mean free path of the particle surface and that the distance is greater than the turnover distance σ (where the radial derivative of the effective potential vanishes) of the effective potential implicitly defined by

$$J^2 = \frac{e_1^2}{2} \frac{\kappa-1}{\kappa+1} \frac{2\sigma^2-1}{(\sigma^2-1)^2} - \frac{e_1 e_2 \sigma}{2} \tag{5.8}$$

where J^2 is the angular momentum of the ion. Note that in no way are the essential aspects of the interaction potential and Hamiltonian orbit (or constant characteristic) of the ion not taken completely into account. The effectiveness of the repulsive Coulomb interaction is substantially modified by the image force only if the ion is on the proper orbit which is determined to allow the effective potential to change signs thereby entering the region where the interaction force is monotonically increasing with decreasing r. Details of these computations were discussed in MARLOW and BROCK [5.9], and MARLOW [5.14].

The results of these computations can be given in the form

$$F = G_0 - \frac{\lambda}{2\sqrt{\pi}} G_1 \quad , \quad G_1 = g + k \tag{5.9}$$

where F is the total ionic flux to a particle, G_0 is the free-molecular flux [5.8], and G_1 is the correction to the free-molecular flux described above. G_1 was determined as the sum of two terms: g is the flux to the sphere determined by $J^2 = 0$ in (5.8), and k is the flux for particles with a tangential component, $J^2 \neq 0$, to the sphere.

Figure 5.1 is a graph illustrating the relative importance of the flux correction term G_1 with respect to the free-molecular flux term G_0. It is graphed for $e_1' = 2$ and $e_1' = 1$ which correspond to 13.9 nm particles with dielectric constants $\kappa = \infty$ and $\kappa = 1.67$, respectively. The argument e_1e_2 is simply a measure of the Coulomb interaction which is independent of κ. For a constant Coulomb potential e_1e_2, an increase in κ corresponds to an increase in the turnover distances. Since collisions with neutrals increase in importance as the capture distance increases, the flux correction G_1 must increase. The fact that, e.g., $(-G_1/G_0)_{e_1=1} < (-G_1/G_0)_{e_1=2}$, but both have the same functional dependence upon e_1e_2, can be interpreted to mean the Coulomb interaction governs the "availability" of ions to be captured while playing a small role in the actual capture rate. Again, as the strength of the Coulomb interaction increases, the importance of the collisional interaction increases. At large Coulomb potentials, the two curves increase at a constant ratio indicating the vanishing in importance of the image force per unit increase in the Coulombic force.

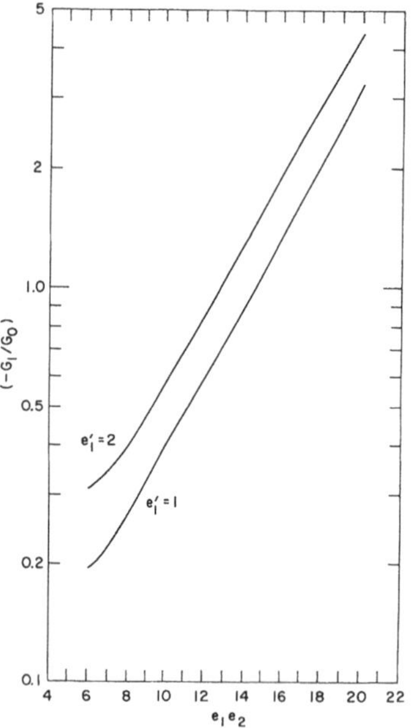

Fig.5.1. Results of model calculation of near-free molecular particle charging: ratio of collisional flux correction G_1 to free molecular flux G_0 vs nondimensionalized Coulomb interaction; e_1' parameterizes image potential (see Sect.5.2.2 of text)

Since these are model calculations based upon near-free-molecular gas densities, the example given may be considered meaningful for pressures under 5 atm and for particles of under 100 nm.

These results have implications for all aerosol interactions. They motivate the detailed attention given in the following sections to matters dependent upon the chemical physics of the separated but interacting particles. At the same time, the kinetic theoretical component of aerosol interactions will not be treated further. Nevertheless, the central role of the gas in fixing the physical boundary conditions must always be accounted for. It determines the dimensions over which details of force structure are important, thereby governing details of the interactions. For example, it is well known [5.15] that despite the importance of the image force in the charging of transition and free-molecular regime particles, it apparently plays no significant role in the continuum regime. The implications this holds for aerosol measurement by electrical charging methods have been pointed out [5.16].

The interaction of spherically symmetric systems can be used as a guide for systems of more complex geometry. The possibility arises that a coupling can develop between a particle (or large, nonspherical molecule) with an orientation-dependent interaction force and a corresponding orientation-dependent Knudsen number which would need to be defined for a particle whose different dimensions were characteristic of, e.g., both transition and continuum regimes. Scientific and technological implications of such behavior could be considerable.

5.3 Electrostatic Interaction Forces

5.3.1 Multipole-Multipole Interactions

As pointed out in the introductory section of this chapter, there are two types of electrostatic interaction forces between neutral bodies: the multipole-multipole (Keesom-type) forces and the induced multipole forces (Debye-Falkenhagen-type).

The electrostatic interaction potential energy for distributions of charges on two bodies can be written in a variety of ways [5.6]. For the purposes of this review the following formula is most illuminating:

$$V \lesssim \sum_{ij} q_i q_j \sum_{ab} \frac{r_i^a r_j^b}{R^{a+b+1}} B_{ab} \sum_{m=-a}^{\alpha} \frac{Y_a^m(\theta_i \phi_i) Y_b^{-m}(\theta_j \phi_j)}{[(a+m)!(a-m)!(b=m)!(b-m)!]^{\frac{1}{2}}} \tag{5.10}$$

with $\alpha = \min(a,b)$, $B_{ab} = (-1)^\alpha [4\pi(a+b)!]/[(2a+1)!(2b+1)!]^{\frac{1}{2}}$. In this inequality, the indices i and j label charges $q_i q_j$ on the first and second particles, r_i and r_j give the distances of charges i and j from the origins of local coordinate systems placed within those bodies, R is the distance between the origins of those coordinate systems, $\theta_i\phi_i$ and $\theta_j\phi_j$ are the local polar coordinates of q_iq_j, and Y_a^m are spherical harmonics. Equation (5.10) is expressed as an inequality because the interaction potential between material media must necessarily include their dielectric constants, thereby reducing their interaction potential. This also means an induced or image interaction potential is present as will be discussed. To estimate the importance of the Keesom potential for aerosol interactions, it must be compared with the particle's thermal energy at a distance of approximately a gas molecular mean free path, ℓ_g, from the body it interacts with. Then

$$V_k(|r_a - r_b| = \ell_g)/kT \gg 1 \tag{5.11}$$

is a necessary, but not sufficient, condition for the importance of V_k as an interaction capable of perturbing the particle's motion and affecting the collision probability. A second condition that places a somewhat more severe limitation upon the effectiveness of Keesom forces in aerosols arises from the fact that the net force averages to zero upon a rotation. Therefore, the particle must rotate slowly by comparison with its translational motion. This is estimated qualitatively by requiring the period for rotational motion Λ to be somewhat greater than the particle's translational relaxation time τ_t. With the particle's corresponding path length for translational relaxation denoted ℓ_p, this condition becomes

$$\frac{\Lambda}{\tau_t} \sim \left(\frac{1}{\ell_p}\right)\left(\frac{I}{m}\right)^{\frac{1}{2}} \gg 1 \tag{5.12}$$

where I is the particle's moment of inertia and m is its mass. From this estimate, it is evident that Keesom forces are unimportant for transition and free-molecular regime particles at pressures within an order of magnitude of atmospheric pressure.

5.3.2 Induced Electrostatic Interactions

Debye-Falkenhagen-type forces do not suffer from the vanishing of the rotationally averaged interaction since they arise from a force due to induction which can track the rotation of the particle. Unlike the Keesom force in which, in principle, the interaction of distributed point charges are considered, induction forces are dependent upon a collective molecular or material property, the static dielectric constant. Definition of the domain of validity for the classical calculation is required and has been given in the case of metals [5.17]. Those authors pointed out that if the surface of the metal were taken at the center of mass of the surface charge distribution, then classical electrostatic calculations would be valid. In most aerosol problems this correction will be unimportant because the actual surface charges are spread over a thickness of about 0.2 nm.

Generally speaking, the interaction forces between two dielectric bodies with multipolar charge distributions are extremely difficult to compute. No attempt to review the literature is made here. Rather, a recent paper which addresses this problem in some generality for sphere-sphere and sphere-plane interactions among dielectrics is cited, and some results indicative of the importance of considering distributed, rather than point, charges and multipoles are described.

BUFF et al. [5.18] gave a general formulation for the interaction energy of a uniform dielectric sphere containing fixed point charges with another sphere of possibly differing dielectric constant and fixed charge distribution. Both spheres are immersed in a dielectric medium. They showed that this energy is the same as that between such a sphere and a plane dielectric if some simple interpretations of the variables were made. For the purposes of aerosol interactions, it is generally not necessary to consider the multipole-multipole contributions to the energy, thereby greatly simplifying the results of this work which are nonetheless quite complicated. Conversely, their contributions in this and a previous paper [5.19] are useful in the computation of the interaction energy between a dielectric particle with a fixed charge distribution and a plane dielectric.

In the case of two identical interacting spheres of dielectric constant k, with charges q_1 and q_2 at their centers, radii b, and intercenter distance R the interaction energy $W_I^{(c)}$ is shown [5.18] to have the fairly accurate asymptotic value

$$W_I^{(c)} = \frac{q_1 q_2}{R}\left[1 + \frac{(q_1+q_2)}{4q_1q_2}\, F_{II}^{(c)}(\lambda = 1) - \frac{(q_1-q_2)^2}{4q_1q_2}\, F_{II}^{(c)}(\lambda = -1)\right] \tag{5.13}$$

where

$$F_{II}^{(c)} \simeq (\eta\omega)^{-1}\left\{\left[\sum_{n=1}^{\infty} (-\eta)^{n-1}\right]\left(\frac{\sinh\beta}{\sinh\eta\beta}\right)^{-1} - 1\right\} - 1 \tag{5.14}$$

$$\eta = \frac{\lambda(\kappa-1)}{\kappa+2}\, {}_2F_1\left(1, \frac{1}{1+\kappa}\ ;\ 2+\frac{1}{1+\kappa}\ ;\ \omega^2\right) \tag{5.15}$$

$$\cosh\beta = \frac{1}{2\omega}\ ;\quad \omega = \frac{b}{R}\quad . \tag{5.16}$$

The interaction energy $W_{II}^{(c)}$ between a single such sphere and a plane of dielectric constant D has the asymptotic value

$$W_{II}^{(c)} = \frac{-q^2}{4b^2}\left[1 + F_{II}^{(c)}(\lambda = -1)\right]\quad . \tag{5.17}$$

The importance of the image potential is indicated in [Ref.5.18, Table IIA]. There, it is shown for example, that the ratio of repulsive forces between two identical dielectric charged spheres separated by 0.041 and 0.389 radii to the point charge Coulomb forces for the same separation are respectively 0.74 and 0.88 if the dielectric constants of the spheres are 5.

For the case of a point dipole in a dielectric sphere interacting with its image, the dipole moment (μ) is expressed in terms of components perpendicular (μ_{pe}) and parallel (μ_{pe}) to the plane of the interface. BUFF et al. [5.18] show the interaction energy to have an asymptotic value

$$W_{II}^{d} \simeq \frac{1-D}{2(1+D)} \frac{1}{(2Z_0)^3} \left(\frac{3}{2+\kappa}\right)^2 \left(\mu^2 + \mu_{pe}^2\right) + \left(\frac{1-D}{1+D}\right)^2 \left[\frac{b^3}{2D(2Z_0)^6}\right] \left(\frac{1-\kappa}{2+\kappa}\right) \left(\frac{3}{2+\kappa}\right)^2 \left(\mu^2 + 3\mu_{pe}^2\right) \tag{5.18}$$

where D is the dielectric constant of the plane dielectric, Z_0 is the distance of the center of the sphere from the planar interface, and k and b are as before. Contrary to the case above of the charged dielectric sphere, (5.17), this approximation is shown not to be particularly accurate for small separations s, where $s = Z_0/b$. When the exact expression was evaluated for the sphere in contact with the plane, the authors found it to be so poor an approximation as not to be useful. This conclusion can be assumed true at close range (as they show for mathematically analogous but physically dissimilar cases to aerosols) where details of the aerosol interaction forces are of greatest importance.

From the foregoing discussion it can be stated that a useful expression exists for calculating the interaction energies among charged dielectric spheres, while the induced multipole interaction is in poor shape without numerical computation.

The question of a point charge moving near a dielectric surface has recently been examined [5.20]. To make the problem tractable, it was solved under the constant-velocity approximation. The author showed that the force acting upon the particle differed depending upon whether it was moving toward or away from the interface and that this force at any instant was dependent upon the particle's past history because of coupling to the surface plasmons. Such investigations have significance for an electron with kinetic energy greater than its binding energy to the surface it interacts with. Because of the necessary energy for a charged cluster to move at a velocity comparable to that for an affected electron ($\sim 10^5$ m/s) this effect is usually of no importance in aerosol considerations.

5.4 Van der Waals Interactions of Microparticles

5.4.1 Background

The aerosol interaction forces discussed in the preceding sections of this chapter have involved only electrically charged particles or neutral particles with multipole moments. The importance of charge on all phases of aerosol dynamics is generally recognized and is exploited in numerous ways. It was argued above (Sect. 5.3.1) that neutral multipolar aerosol particles are unlikely to be of general importance in aerosol processes for fundamental reasons connected with their formation. This leaves only the short-range forces involving neutral, nonpolar particles, the van der Waals forces. Since they bear high chemical and structural specificity, there is reason to believe they may be quite important in ways that are customarily overlooked. For example, ZAGAINOV et al. [5.21] have done experiments showing that the efficiency of removal of Fe_2O_3 molecular clusters from an aerosol upon passage through 300 mm long by 8 mm i.d. tubes of differing materials vary substantially depending upon the tube material. Since composition-independent removal of particles is the basis of operation of the diffusion battery that is widely used for aerosol size distribution measurements [5.22] and the diffusion processor used to obtain size related separations among particles for subsequent chemical analysis [5.23], measurement procedures may require reexamination for their general validity. Similarly, the modeling of coagulation that is so fundamental to the understanding of all aerosol processes [5.24] must be considerably altered if certain similar or dissimilar particles cannot coagulate, their coagulation is moderated by unrecognized environmental conditions, or they adhere too weakly to be stable. Conversely, the interaction of molecules with particles is also a chemically specific process. Modern van der Waals theory can not only give important insight into the interaction potential the molecule experiences in the vicinity of the particle, which determines collision frequency as discussed in Sect.5.2 above, but also in many cases can help one to understand the actual chemical physics of molecule-particle interaction. The breadth of applicability of modern van der Waals theory arises from its basis in the measurable and microscopically interpretable quantities, the electric and magnetic susceptibilities of the interacting media.

The domain of importance of the van der Waals interaction for aerosol microphysics is typically for particle separations under a radius. For example, consider two identical spherical particles at room temperature with $kT \cong 4.14 \times 10^{-14}$ ergs. Assume the particles' electric susceptibilities $\varepsilon(\lambda) = 2$ for 400 nm $< \lambda <$ 700 nm and $\varepsilon = 1$ for all other wavelengths. Then if the particles are separated by a distance of a radius, their van der Waals interaction energy is on the order of 1.5×10^{-15} ergs, whereas it is 2×10^{-14} ergs when the separation is 0.3 radius and 1.2×10^{-13} ergs at a separation of 0.1 radius. By inclusion of only a narrow range of interaction wavelengths, the absolute magnitude may be low; however, the relative magnitudes explain the success of Fuchs' dividing sphere approximation and aerosol macrophysical coagulation models which assume contact interaction only: thermal motion dominates a particle's transport until it is very close to another particle.

Van der Waals theory for material interactions has developed starting from two complementary paths. One has been the approach of describing, by various means, the perturbation to the electromagnetic field in the region between two bodies due to their presence. The interaction energy is then due to the difference in field energy between its values with and without the presence of the bodies. The other path is the summation of individual interatomic interactions between the two bodies taking into account their presence in condensed media. Another very important method derivative of the first examines the surface plasmon modes of the interacting particles. Since these are necessarily coupled to the external fields, their variation with distance of the interacting bodies is related to the interaction force.

A conceptually straightforward, though algebraically lengthy, formulation of the interaction summation method has been given by LANGBEIN [5.25] in which he employs the Drude model for the elementary constituents of the interacting bodies. Each of these constituents (molecules here) is a dipole oscillator of amplitude u_i and unperturbed frequency ω_i which interacts with all other molecules via their dipole fields. The Hamiltonian for this interacting dipole oscillator system is then

$$H = \frac{m}{2} \sum_i \left(\dot{u}_i^2 + \omega_i u_i^2\right) - \frac{e^2}{2} \sum_{ij} u_i T_{ij} u_j \tag{5.19}$$

where T_{ij} is the dipole field tensor, $T_{ij} = -\nabla_i \nabla_j (1/|r_i - r_j|)$, that appears in the usual multipole expansion of the electric potential (5.10). Here r_i and r_j are the positions of distinct molecules and $T_{ii} = 0$. Application of Hamilton's formalism and standard perturbation theory techniques gives an expression for the new frequencies Ω_i of the system which now includes the oscillator dipole interactions. The dispersion energy ΔE of this system is the difference between the perturbed and unperturbed energies, $\Delta E = (\hbar/2) \sum_i (\Omega_i - \omega_i)$, and LANGBEIN shows that it can be written

$$\Delta E = \frac{\hbar}{4} \left[\left(\frac{e^2}{m}\right)^2 \Sigma_{ij} T_{ij} T_{ji} \frac{1}{2} \Sigma_2 + \left(\frac{e^2}{m}\right)^3 \sum_{ijk} T_{ij} T_{jk} T_{ki} \frac{1}{3} \Sigma_3\right.$$

$\hbar = h/2\pi$ (normalized Planck's constant)

$$+\left(\frac{e^2}{m}\right)^4 \sum_{ijk\ell} T_{ij}T_{jk}T_{k\ell}T_{\ell i}\frac{1}{4}\Sigma_4 + \ldots\Big] \qquad (5.20)$$

where

$$\sum_n = \sum_n (\omega_1,\omega_2, \ldots, \omega_n) = \frac{1}{\pi}\int_{-\infty}^{\infty} d\omega \prod_{i=1}^{n} \frac{1}{\omega^2+\omega_i^2} \quad . \qquad (5.21)$$

Equation (5.20) can be interpreted to mean that the dispersion energy ΔE is composed of all the combinations of interactions (disregarding order) among n units where the energy of each interactive combination of P units is $-(\hbar/4)(e^2/m)^P \sum_{ij}(T_{ij}T_{jk}\ldots)(1/P)\Sigma_P$.

Now consider the (multipole) eigenfrequencies of each molecule j. If the sums in each term of (5.20) be rearranged in recognition of the fact that

$$\alpha_j(i\omega) = \frac{e^2}{m}\sum_{\substack{n \\ (n\,\mathrm{at}\,j)}} \frac{1}{\omega^2+\omega_n^2} \qquad (5.22)$$

forms a physically meaningful combination, the molecular polarizability in the Drude model, (5.20) becomes

$$E = -\frac{\hbar}{4\pi}\int_{-\infty}^{\infty} d\omega \left(\frac{1}{2}\sum_{ij} \alpha_i T_{ij}\alpha_j T_{ji} + \frac{1}{3}\sum_{ijk} \alpha_i T_{ij}\alpha_j T_{jk}\alpha_k T_{ki} + \frac{1}{4}\sum_{ijk\ell} \alpha_i T_{ij}\alpha_j T_{jk}\alpha_k T_{k\ell}\alpha_\ell T_{\ell i} + \ldots\right) \quad . \qquad (5.23)$$

This is an expression for the dispersion energy of a body in terms of an independent measurable property of its constituents. It shows that the dispersion interaction among molecules in one body generally cannot be approximated by the pairwise interactions among them [first term in (5.23)], but must include all combinations of interactions. Formulation of the two-body dispersion interaction is now clear. Suppose that a dipolse j is located externally to a body A and its interaction with a molecule i in A is sought. That interaction is given by the sum of the dipole fields due to the molecule alone and to its iterated multiple interactions with all the other molecules in A. The resultant dipole field

$$T_{ij}^{SCR} = T_{ij} + \sum_{k\in A} T_{ik}\alpha_k T_{kj} + \sum_{k,m\in A} T_{ik}\alpha_k T_{km}\alpha_m T_{mj} + \ldots \qquad (5.24)$$

is called the screened dipole field at j due to the molecule i that is part of body A. If j is part of a body B, the interaction energy of A with B is then due to the iterated sum of interactions of the screened fields of A with B.

$$\Delta E_{AB} = -\frac{\hbar}{4\pi}\int_{-\infty}^{\infty} d\omega \Big(\sum_{i\in A}\sum_{j\in B} \alpha_i T_{ij}^{SCR}\alpha_j T_{ji}^{SCR}$$

$$+\frac{1}{2}\sum_{i,k\in A}\sum_{i,\ell\in B} \alpha_i T_{ij}^{SCR}\alpha_j T_{jk}^{SCR}\alpha_k T_{k\ell}^{SCR}\alpha_\ell T_{\ell i}^{SCR} \tag{5.25}$$

$$+\frac{1}{3}\sum_{i,k,m\in A}\sum_{j,\ell,n\in B} \alpha_i T_{ij}^{SCR}\alpha_j T_{jk}^{SCR}\alpha_k T_{k\ell}^{SCR}\alpha_\ell T_{\ell m}^{SCR}\alpha_m T_{mn}^{SCR}\alpha_n T_{n\ell}^{SCR} + \dots\Big) \quad .$$

This is the dispersion energy of A due to the presence of B and its physical interpretation is that it arises from the multiple "reflected" interactions of the screened fields at each point in A with the oscillators of B as characterized by their own screened fields.

An explicit representation of the interaction energy for homogeneous half-spaces can be derived by substitution of physical variables and use of the Clausius-Mosotti formula connecting molecular electric susceptibility α_j with the material dielectric constants ε_1 and ε_2 of the two half-spaces. Then the interaction energy per unit area of two half spaces separated by a distance d such that $d < (2\pi c/\omega_{min})$, c = speed of light and $\omega_{min}(\omega_1,\omega_2, \dots)$ is given by

$$\Delta E_{AB} = -\frac{\hbar}{16\pi^2 d^2}\int_0^{\infty} d\omega \sum_{n=1}^{\infty} \frac{1}{n^3}\left[\frac{\varepsilon_1(i\omega)-1}{\varepsilon_1(i\omega)+1}\,\frac{\varepsilon_2(i\omega)-1}{\varepsilon_2(i\omega)+1}\right]^n \quad . \tag{5.26}$$

The nonretardation condition on the separation d is implicit in the derivation since no accounting of the exchange photon's propagation is explicitly included. ISRAELACHVILI [5.26] has given an alternative, physically intuitive treatment of van der Waals interactions based upon interactions of images across media of differing dielectric susceptibilities. This treatment represents a generalization of aspects of an approach originated by McLACHLAN [5.27-30]. Though the work of McLACHLAN nominally treats forces on molecules, it contains the germ of the future developments which facilitated practical application of Lifshitz theory.

The other approach to van der Waals forces which focuses upon the interaction field between the bodies originated in the works of CASIMIR and POLDER. However, it was the seminal paper of LIFSHITZ [5.31] which demonstrated its general applicability and its origins in the spontaneous quantum-mechanical dipole fluctutations of the elementary constituents of matter. The original derivation of (5.26) by LIFSHITZ is described here as an introduction to the kinds of consideration of this method.

In dielectric nonmagnetic media, Maxwell's equations may be written, for monochromatic fields,

$$\nabla \times E = i \frac{\omega}{c} H$$
$$\nabla \times H = - i \frac{\omega}{c} (\varepsilon E + K) \qquad (5.27)$$

where K is the polarization field due to the random fluctuations of the medium. Due to the randomness of the fields, the correlation function $<K_i(\underline{r}_i)K_j(\underline{r}_j)>$ is argued to be

$$<K_i(\underline{r}_i)K_j(\underline{r}_j)> \propto \delta_{ij}\delta(\underline{r}_i - \underline{r}_j) \quad . \qquad (5.28)$$

Then the fluctuation-dissipation theorem [5.32] gives an expression for the constant of proportionality, the mean square value of K. Equation (5.27) was solved by LIFSHITZ for E and H in half-spaces of media 1 and 2 applying the appropriate boundary conditions. In the vacuum between the half-spaces, the solution is given by the homogeneous versions of (5.27) with the required boundary conditions at the two surfaces. These solutions are then used in the Maxwell stress tensor, with all frequencies being integrated over, to give the forces acting upon the surfaces of the two half-spaces. The result is (5.26) in the nonretarded limit, that is, when the separation is less than the wavelength of the least-energetic optical excitation. In the original paper, the fluctuation-dissipation theorem is not cited explicitly but is at the foundation of the method used. In fact, it is the basis for modern van der Waals theory which frequently employs it and its generalization, linear response theory (see [5.33] and references therein). Equation (5.28) is essential in this formulation because it "localizes" the perturbation to the system. That is, regardless of the mechanism of its activation, once the perturbation occurs, the dissipation of energy (radiation here) within the medium is rigorously characterizable as local. This is not necessarily true in all media, and modifications of the theory are required where spacial dispersion can occur, as in metals and electrolytes. Locality is also implicitly invoked in the first method described where individual interactions were summed over. In particular, the fundamental interacting unit between the two bodies is T_{ij}^{SCR} of (5.24) and it alone describes the two-body interaction in (5.25).

Historically, van der Waals forces between condensed media have been calculated by summing over pairwise interactions between the molecular constituents of each body [e.g., the first term in (5.23)]. This method, called Hamaker theory, first showed that the sum of the intermolecular interactions, as calculated by elementary second-order perturbation theory in the Schrödinger equation, between macroscopic bodies gives realistic attractive forces in many cases. Its description can be found in any reference on modern dispersion or van der Waals forces between condensed media including LANGBEIN [5.34] and MAHANTY and NINHAM [5.35] at the advanced level or the excellent elementary introduction by PARSEGIAN [5.36].

Hamaker theory is the classical theory of van der Waals forces and is what is generally invoked even today in many fields including aerosols, despite its long-recognized inadequacies. Briefly, these inadequacies are as follows [5.35]:

1) failure to account for many-body and collective properties of the medium [e.g., in (5.24), omission of all terms with summations],
2) failure to incorporate a frequency-dependent polarizability as a basic component (5.26) describing the interaction,
3) inability to handle spurious divergences in the theory due to assumption of point molecules,
4) inability to accommodate plasticity of interacting particles upon close approach.

YASUDA [5.37] calculated the interaction forces between graphite masses by the Lifshitz method and by pairwise additivity (Hamaker method) and found substantial differences. He considered two graphite half spaces 10 nm apart with their lattice planes parallel to the interfaces. By using literature experimental data for the imaginary part of graphite's dielectric constant and the Kramers-Kronig relation for $\varepsilon(i\omega)$, YASUDA calculated the force of attraction according to Lifshitz theory $F_L(10\ \mathrm{nm}) = 4.76\times 10^4\ \mathrm{dyn/cm^2}$. Under the assumption of pairwise additivity, he found $F_S(10\ \mathrm{nm}) = 20.86\times 10^4\ \mathrm{dyn/cm^2}$ as the attractive force. The discrepancy was largely due to shielding effects. Other discussions of Lifshitz vs Hamaker theory were given by PARSEGIAN and NINHAM [5.38] and SMITH et al. [5.39]. These papers discuss the role of intervening media between the interacting bodies.

5.4.2 Computational Approaches

State Density Integration and Retardation

The interaction energy of two particles at $\underline{R}_1$ and $\underline{R}_2$ with tensor electric susceptibilities $\alpha_{uv}^{(1)}(\omega)$, $\alpha_{uv}^{(2)}(\omega)$ that are functions of frequency ω may be determined by computing the effect of their presence upon the eigenfrequencies of the electromagnetic field. Using these susceptibilities in the suitably Fourier-transformed equation relating induced polarization $\underline{P}$ and external electric field $\underline{E}$ gives

$$P_u(\underline{r},\omega) = \sum_{v=1}^{3} \alpha_{uv}(\omega)E_v(\underline{R},\omega)\delta(\underline{r} - \underline{R}) \quad ,$$

where $\underline{r}$ is the observation point. The polarization current $j_u = (\partial/\partial t)P_u(r,t)$, where $P_u(r,t)$ is the Fourier transform of $P_u(r,\omega)$, is then used as the source term in Maxwell's equations. Determination of the interaction energy is accomplished most easily by working in the Lorentz gauge and Fourier transforming the equations. From this procedure the following integral equation solution is derived:

$$E_u(\underline{r},\omega) = -4\pi \sum_{wv}\sum \left[\alpha^{(1)}_{uv}(\omega) G_{vw}(\underline{r},\underline{R};\omega) E_w(\underline{R}_1,\omega) + \alpha^{(2)}_{uv}(\omega) G_{vw}(\underline{r},\underline{R}_2;\omega) E_w(\underline{R}_2,\omega)\right] \quad . \tag{5.29}$$

Here, $G_{vw}(\underline{r},\underline{R}_{jw})$ are the Green's functions for the full field equations. By letting $\underline{r} = \underline{R}_1$ and $\underline{r} = \underline{R}_2$, a pair of coupled equations for the electric fields at the two point particles is derived. The roots of the determinant $D_{12}(\omega)$ of their coefficients give the eigenfrequencies of the completely coupled system (two particles and interacting field). If $D_0(\omega)$, $D_1(\omega)$, and $D_2(\omega)$ are respectively the similarly derived determinants for the free field and for the field coupled only to particles 1 and 2, then the roots of the entire system, $D_{12}(\omega) = 0$, $D_0(\omega) = 0$, $D_1(\omega) = 0$, $D_2(\omega) = 0$ suffice to express the zero-point energy shift E_{12} of the interacting system from the fully uncoupled system:

$$\Delta E_{12} = (\hbar/2)\left\{\sum \left[R[D_{12}(\omega)] - R[D_0(\omega)]\right] - \sum \left[R[D_1(\omega)] - R[D_0(\omega)]\right] - \sum \left[R[D_2(\omega)] - R[D_0(\omega)]\right]\right\} \tag{5.30}$$

where $R(X)$ = roots of the equation $X = 0$. Since

$$\sum \left[R[X(\omega)] - R[Y(\omega)]\right] = \frac{1}{2\pi i} \oint \omega \frac{d}{d\omega} \ln \left[\frac{X(\omega)}{Y(\omega)}\right] d\omega \quad , \tag{5.31}$$

$$\Delta E_{12}(R_{12}) = \frac{1}{2\pi i} \oint d\omega \frac{d}{d\omega} \ln \left|\frac{D_{12}(\omega)/D_0(\omega)}{[D_1(\omega)/D_0(\omega)][D_2(\omega)/D_0(\omega)]}\right| \quad , \tag{5.32}$$

where the contour includes the positive real axis. This form of the interaction energy, especially as it is derived from (5.30), is often used in van der Waals theory and provides an extremely powerful and general technique for treating the interaction energy of systems [5.34,35]. By expanding (5.32), retaining terms of up to second order in the α's, and assuming them to be isotropic, the interaction energy for point particles can be written

$$\Delta E_{12}(R_{12}) \cong \frac{-\hbar}{\pi R_{12}^2} \int_0^\infty d\xi \alpha^{(1)}(i\xi)\alpha^{(2)}(i\xi) \exp(-2\xi R_{12}/c) \times \left(\frac{\xi^4}{c^4} + \frac{2\xi^3}{c^3 R_{12}} + \frac{5\xi^2}{c^2 R_{12}^2} + \frac{6\xi}{c R_{12}^3} + \frac{3}{R_{12}^4}\right) \tag{5.33}$$

where $\xi = i\omega$. *If* $R_{12} \ll c/\omega$, *the nonretarded region,*

$$\Delta E_{12}(R_{12}) \simeq -\frac{3\hbar}{\pi R_{12}^{6}}\int_0^\infty d\xi\alpha^{(1)}(i\xi)\alpha^{(2)}(i\xi) \quad (5.34)$$

which is the London formula. *If* $R_{12} >> c/\omega$, *the retarded region,*

$$\Delta E_{12}(R_{12}) = -\frac{23\hbar c\alpha^{(1)}(0)\alpha^{(2)}(0)}{4\pi R_{12}^{7}} \quad (5.35)$$

which is the Casimir and Polder result [5.40]. In (5.34,35) the separation dependence of the force law is explicit. In that region where the interparticle separation is considerably less than the wavelengths where the susceptibilities are at their maxima, an R_{12}^{-6} law prevails, whereas an R_{12}^{-7} law holds where the wavelength exceeds the separation. In the limit for this latter case, only static fields contribute as illustrated in (5.35). Since the locations of the maxima in the susceptibilities are dependent upon composition there is no universal interparticle interaction force law. MAHANTY and NINHAM [5.35,41] may be referred to for a full exposition of the approach and MAHANTY and NINHAM [5.42] showed how it could be applied to extended particles.

Surface-Mode Hypothesis

VAN KAMPEN et al. [5.49] proposed that the interaction forces for macroscopic media could be calculated by considering only the surface-mode solutions of Maxwell's equations at all interfaces. This method was extended [5.44,45] and is used in various applications. For the original case of two half-spaces separated by a gap d, the solutions are sought to the equations of electrostatics subject to the conditions $\underline{\nabla} \cdot \underline{D} = 0$ and $\underline{\nabla} \times \underline{E} = 0$ with no spacial dielectric dispersion, but with boundary conditions at the interfaces. By matching boundary conditions and requiring vanishing solutions at infinity, the dispersion relation

$$D(\omega) \equiv \frac{1}{4}\{[\varepsilon(\omega)+1]^2 - [\varepsilon(\omega)-1]^2\exp(-2kd)\} = 0 \quad (5.36)$$

for the surface-mode frequencies is derived. Application of the procedure of the preceding paragraph, (5.30,31), provides an interaction energy corresponding to Lifshitz's results [e.g., (5.26)]. Choice of boundary conditions and the consequent determination of the dispersion function [e.g., (5.36)] are particularly important in this approach, not only as a computational procedure, but also for the detailed accounting of the physics of the interaction. The surface modes involved are plasmons and their full description is a study of its own.

A particularly lucid treatment and comparison of the surface-mode and Lifshitz approaches to the calculation of the van der Waals interaction forces was given by

NIJBOER and RENNE [5.46] in which they concluded that the former method was useful only when there was no damping [$\mathrm{Im}\{E(\omega) = 0\}$] of the surface modes. SCHRAM [5.47] and LANGBEIN [5.48] questioned the general validity of the surface-mode approach because of surface damping and of the lack of proper definition of the integration domains in the energy expression following from (5.31). These difficulties lie in the determination of the zero-point energies of the systems which are subtracted from the total energy of the interacting system to get the interaction energy. Langbein analyzed both the Lifshitz and surface-mode approaches in terms of analytic structure of the dispersion functions that may be shown to enter in both cases. He interpreted the structure of the appropriate forms of (5.31) in terms of the vacuum modes of the electromagnetic field and found the conceptual difficulties encountered in the original approaches [5.31,43] to be traceable to an inadequate accounting of these vacuum modes. This problem was removed by first solving the problems with finite boundary conditions, wherein all modes not coupled to the system and, therefore, irrelevant are excluded. Subsequently, the boundaries are allowed to go to infinity. DAVIES [5.49] determined that by formulating the electromagnetic free energy in terms that eliminate nonlinear processes (in agreement with the fluctuation-dissipation theorem), conditions for the general validity of the surface mode approach can be unambiguously laid down. Specifically, they require the following of the dispersion function $D(\omega)$: 1) the zeroes of D determine the normal modes of the system; 2) D is meromorphic in ω and the poles are independent of configuration perturbations; 3) $\ln[D_1(\omega)/D_0(\omega)]$ where D_1 and D_0 are different nearby configurations, does not grow faster than exponentially in the lower half-plane and goes to zero for large ω in the upper half-plane and on the real axis. With these guidelines, it is possible to construct the dispersion relations in a fashion that avoids the analyticity difficulties (i.e., by judicious use of Green's functions) discussed by LANGBEIN while remaining mathematically and physically correct.

5.4.3 Modeling the Dielectric Susceptibilities

In aerosol van der Waals forces, the central role played by the material properties of both the particle and whatever it interacts with is clearly illustrated in the above considerations where the polarizability or electric susceptibility α and the dielectric constant (or permeability) ε are used alternatively. Provided that α represents the material's susceptibility, not the molecular polarizability, the usual definition

$$\varepsilon(\omega) = 1 + 4\pi\alpha(\omega) \tag{5.37}$$

pertains. The value of this identification is that ε is experimentally accessible for any system of homogeneous but unknown composition. For composite particles of known constituents, literature values of ε may be used. This permits this relatively simple composition-dependent physical parameter to determine the aerosol interaction forces for separations S such that 500 nm > S > 0.5 nm.

Among the ways to determine the dielectric constant as a function of frequency, the most portentious for aerosol work is by the use of reflection and transmission data for electromagnetic waves. Through reflectance measurements, the refractive index and dielectric constant are determined via Kramers-Kronig analysis [Ref.5.50, pp.256-266] of the data [5.51-53] for plane surfaces. Transmission measurements are done on powder samples [5.54] and matrix-isolated samples [5.12] as well as by other means. KREIBIG [5.55] has shown how to extract the complex dielectric constant from absorption data. Neither technique requires large quantities of material and at the same time they implicitly include the surface of the material in contact with air or a vacuum. The powder and matrix methods are especially promising for aerosol work since they may be useful in the dielectric characterization of ill-defined particulate matter sampled directly from its source. DERJAGUIN et al. [5.56] have pointed out that the imaginary part of the dielectric susceptibility, which is related to optical absorption, is all that is needed for the calculation of molecular van der Waals forces according to Lifshitz theory.

NINHAM and PARSEGIAN [5.57] and PARSEGIAN [5.36] have given a phenomenological parameterization for $\varepsilon(\omega)$ whose form is based upon the known response of any materials in each frequency interval from a constant field through the ultraviolet. By using experimentally determined data at representative points of the spectrum, they found that complete information on the spectral response of materials is probably unnecessary, at least when considering component substances which were of similar weight density. Any form chosen for $\varepsilon(\omega)$ must conform to the mathematical constraints on it as a complex function of a complex variable which were discussed in LANDAU and LIFSHITZ [5.50] and elsewhere. With these constraints in mind and with the complex frequency $\omega = \omega_R + i\xi$ they wrote

$$\varepsilon(\omega) = 1 + \frac{C_{mw}}{1+i\omega/\omega_{mw}} + \sum_j \frac{C_j}{1-(\omega/\omega_j)^2 + i\gamma_j\omega} \tag{5.38}$$

for microwave through ultraviolet frequencies, and they used the limiting form

$$\varepsilon(\omega) = 1 - \omega_p^2/\omega^2 \quad , \tag{5.39}$$

where $\omega_p^2 = 4\pi Ne^2/m$, for higher frequencies. In ω_p, the plasma frequency, m, e, and N were respectively the electronic mass, charge, and number density. They used the fact pointed out by DERJAGUIN et al. [5.56] that it is frequently desirable to transform the frequency integration which originally lay along the real axis to an integral along the imaginary axis. There, ε assumes only real values which can be determined via Kramer-Kronig analysis of absorption data, as discussed above.

The first term in (5.38) is for "Debye" relaxation, the orientational response of a polar molecule to an external field. The next term is for Lorentz dispersion which describes the energy dissipation of the damped-harmonic-oscillator model for bound atomic and molecular electrons. The high frequency dielectric permeability is described in terms of the response of all the electrons in the medium being considered as free. By using data for water and hydrocarbons in (5.38,39), NINHAM and PARSEGIAN [5.57] computed the van der Waals forces in lipid-water systems from the generalization of the Lifshitz formula of DZYALOSHINSKI et al. [5.58]. Their results are useful for all considerations of van der Waals interactions, though their form for $\varepsilon(\omega)$ may be inadequate for certain cases discussed below. They find that in the infrared, use of a detailed spectrum for water including three distinct absorption frequencies rather than a single average frequency makes little difference on the interaction. Likewise, as long as the high frequency contributions for the two materials are chosen in a consistent manner, the interaction is relatively insensitive to the choice of frequencies. NINHAM and PARSEGIAN find that as the separation increases, the higher frequencies cease to make an effective contribution to the interaction. Contrary to assumptions used in Hamaker theory, the lower frequencies play a major role so that as the separation increases their role increases due to the effect of "retardation damping" upon the shorter wavelengths. Roughly speaking, as the separation becomes larger than the wavelength of an exchanged photon, that photon ceases to make an effective contribution to the interaction in comparison to longer wavelengths.

Various difficulties with the form of $\varepsilon(\omega)$ or the procedure used by NINHAM and PARSEGIAN may arise when conductors or very small particles are examined. In the first case, nonlocal dispersion has not been adequately treated in van der Waals theory in general, but even if the nonlocal effects could be ignored, interband transitions may need to be accommodated. LANDAU and LIFSHITZ [5.50] propose the form $\varepsilon(\omega) = 4\pi i\omega/\sigma$, where σ is the conductivity, for the very low-frequency dielectric permeability of conductors. Small particles and clusters also must be treated with caution if they are metallic due to surface-scattering and size quantization effects [5.59].

5.4.4 Experimental Corroboration of Lifshitz Theory

The modern formulation for van der Waals forces as originally given by LIFSHITZ [5.31] has been corroborated in a number of experiments. The first of these was described by DERJAGUIN et al. [5.56]. In their measurements, the attractive force between a plate and a spherical lens, both of fused quartz, were determined for distances of separation of 100-1000 nm by use of a beam balance method. The data from this experiment definitively established that the summation of individual molecular interactions method using either the London or Casimir and Polder interactions is inadequate to explain the data, while the Lifshitz method in the retarded region does quite well. Numerous other experiments also operative primarily in the retarded region of separation ($\gtrsim$ 100 nm) have appeared since but are not reviewed here.

The most impressive confirmation of the theory comes from two sets of experiments reported in the early 1970s [5.60,61]. In both cases, determinations of the van der Waals forces covering both nonretarded and retarded distances of separation were made. The significance of such measurements is that they checked the calculations on a single system at all separations, thereby establishing the full theory rather than an asymptotic form of it.

In the first experiment, ISRAELACHVILI and TABOR [5.60] employed a "jump method" in the 1.5-20.0 nm range and a "resonance method" in the 10-130 nm range. Two cleaved mica sheets were glued to glass cylinders which were then mounted opposite each other at right angles on the apparatus. One was on a cantilever leaf spring and the other on a moveable support. When the two approached each other closely enough, the spring-mounted sample jumped into contact with the moveable sample. By monitoring the distance at which this oscurred and knowing the spring constant, force as a function of distance was determined. The resonance method similarly employed a spring, but monitored the change in its resonant frequency as the two mica sheets approached and interacted via van der Waals forces. ISRAELACHVILI and TABOR gave a parameterization of the force law for crossed cylinders as $F = AR/6D^2$ for nonretarded forces and $F = 2\ BR/3D^3$ for retarded forces where D is the separation. If the force law is represented as D^{-n} in these formulas, ISRAELACHVILI and TABOR concluded that in the 2-12 nm range "the force is completely nonretarded with $A = (1.35 \pm 0.15) \times 10^{-19}$ J, $n = 2.0 \pm 0.1$ ... . For separations greater than 12 nm the power law increases above 2.0 and by 50 nm has reached 2.9 ... . Above 50 nm the force is retarded with $B = (0.97 \pm 0.06) \times 10^{-28}$ Jm, $n = 3.0 \pm 0.1$" in complete agreement with Lifshitz theory.

SABISKY and ANDERSON 5.61 used an acoustic method to monitor the thickness as a function of height of a liquid helium film adhering to a SrF_2 cleaved surface. They found complete agreement between their data and Lifshitz theory over the entire range of thicknesses 1-25 nm they measured.

SHIH [5.62] and SHIH and PARAIGIAN [5.63] measured and discussed the van der Waals forces between alkali atoms and a vacuum-deposited gold surface. Though their measurements can be best explained by a $1/R^3$ potential following Lifshitz theory, the agreement is not perfect. Since a vacuum-deposited metal was used, complications attributable to the likely granularity of the gold deposit [5.64] may have arisen. These would include deviations of the gold's electric susceptibility from its bulk value due to size effects and uncertainties in both film thickness and roughness.

ISRAELACHVILI and TABOR [5.65] provide a further review of the experimental literature (as well as of van der Waals forces in general) and DONNERS et al. [5.66] give additional references to other reviews and experimental work.

5.4.5 Constitutive Effects

Charge Carriers

The presence of free charge carriers in either the interacting media or their intervening medium implies nonlocality of the energy dissipation thereby necessitating the reformulation of aspects of the theory. In the aerosol milieu, gas-phase charge carriers are generally of insufficient number densities to play a role in neutral-particle van der Waals interaction forces. Under those conditions where they are numerous, as in aerosol chargers or a weak plasma, the charges carried by the particles themselves dominate the interactions. Therefore, the focus here is upon neutral metallic and electrolytic particles and surfaces, not the intervening gas, and complications they present in the description of their interaction forces.

The simplest van der Waals forces involving free charge carriers occur where only a single substance has the free charges and no other substance in the system being considered can make an electrostatic contribution. For example, an electrolyte sphere coated with a nonpolar hydrocarbon near a pure water aerosol particle is such a system. In this case, the two-particle interaction force can be computed by use of local dielectric permeabilities whenever the charge carrier's plasma frequency is less than the lowest absorption frequency of the system [e.g., ω_{mw} in (5.38)]. See MAHANTY and NINHAM [5.35] for further discussion.

Nonlocality is due to charge carrier interaction with its surroundings. In the "linear" case this is caused by the simple scattering either of conduction electrons from the lattice or surface in metals and other crystalline conductors or of the ion from the solvent molecules in electrolytes. Nonlinear processes may also occur which involve, for example, electron-hole and electron-electron scattering in solids or ion-proton scattering and double layers in electrolytes. In all of these cases the characterization of the dielectric permeability can no longer be expressed solely by expressions like (5.39,40). In practice, the understanding of the dielectric behavior of conductive media is an active research field, the review of which is beyond the scope of this work. The following comments are intended only as one perspective on the relative importance of recent work for understanding the interactions of neutral conductive particles.

Characterization of the electronic responses (and therefore electronic structure) of metallic surfaces is a component of surface physics which is also concerned with the surface energies of metals. Consequently, the methods of the field are appropriate in the detailed description of van der Waals forces among metals. CRAIG [5.67] wrote an influential paper attempting to generalize Lifshitz's approach. HEINRICHS [5.68] pointed to an inconsistency in that paper and carried Craig's methods to the point of calculating the interaction force between two similar slabs of metal separated by a dielectric slab. However, his results could not be generalized to dissimilar slabs. INGLESFIELD and WIKBORG [5.69] used similar methods and HARRIS and JONES [5.70] calculated different quantities. These authors find that for separations greater than about 1 nm their interaction forces become essentially the same as in Lifshitz theory. Despite the fact that they all used the "infinite-barrier" model for the electron potential, whose validity has been demonstrated to be questionable at distances characteristic of absorption, these difficulties evidently cease to be important beyond 1 nm [5.71]. By defining a reference plane Z_0 for a model metal surface in terms of its nonlocal dielectric response function, ZAREMBA and KOHN [5.72] showed how this nonlocality could be included in a calculation of the van der Waals interaction of an atom physisorbed on a metallic surface. It would appear that since the possible variation of Z_0 is small compared with distances of interest here and their result asymptotically converges to the conventional local Lifshitz result at large distances, Zaremba and Kohn's definition of Z_0 can be used to account for nonlocality at small separations while becoming insignificant for large separations. This approach should also be compared with CHAN and RICHMOND [5.73] who used an infinite potential but carried their calculation through to the retarded region by using a "hydrodynamic" model to obtain the permittivities. These results suggest that the work of KREIBIG and co-workers [5.59], who fit permittivities employing infinite surface potentials to experimental data for metallic clusters, may be used successfully in calculating those particles' van der Waals interactions beyond 1 nm.

Investigations of van der Waals forces among electrolytes have been carried out largely in the context of double-layer theory and in connection with phenomena of biological interest [5.74]. The crucial difference between that case and aerosols is the presence of an electrolytic surrounding medium which supports the diffuse portion of the double layer external to the particle. There are a diversity of possible interactions [5.75] that may contribute to electrolytic van der Waals forces many of which are again nonlocal. Since similar arguments for the distance dependence of electrolyte van der Waals forces upon nonlocal dielectric response as pertains in the case of metals may well not apply to electrolytes, a new formulation of the general theory is needed. Hubbard and Onsager's discussion may be very useful in that direction. DAVIES and NINHAM [5.76] consider the interaction of dilute electrolytes in the hydrodynamic approximation for the conduction process. By linearizing the equations they obtain solutions for electrolytic slabs near equilibrium in the nonretarded limit. Further work involving fewer physical assumptions has been published by BARNES and DAVIES [5.77].

An alternative and computationally simpler approach to that of DAVIES and NINHAM [5.76] has been used by PARSEGIAN and NINHAM (unpublished). It is based upon the surface mode hypothesis (see above) modified by replacing the Laplace equation for the electrostatic potential by the Poisson equation which now incorporates the local density of charge carriers. By employing overall electroneutrality and a Debye-Huckel theory expansion, the starting equation $\nabla^2\phi_0 = \Gamma^2\phi$ results. Here

$\Gamma^2 = 8\pi ne^2/\varepsilon kT$ where $n = (1/2)\sum_\nu \nu^2 n_\nu$ and n_ν is the density of ionic species of valence ν. The required dispersion relations are then derived by use of the appropriate boundary and asymptotic conditions upon the solutions ϕ_0. PARSEGIAN and NINHAM summarize their results by pointing out that the effect of mobile ions is incorporated into all results of interactions across single films and multiple layers by replacing each $\varepsilon_k(\omega)$ by $\varepsilon_k(\omega)q_k$ and each distance $q\ell_k$ by $q_k\ell_k$. Here, $q_k^2 = q^2 + \Gamma_k^2$ with q the magnitude of the radial wave vector.

Magnetic Susceptibility

The role of magnetic susceptibility in van der Waals forces is usually ignored due to the weakness of most materials' couplings to magnetic fields relative to their couplings to electric fields. As a result, the theory is substantially less well explored than for the much more important electrical interactions. Nevertheless, there are indications from theory applicable solely to completely retarded interactions that quite important deviations from what is expected for materials of negligible magnetic susceptibility may occur for some substances which have high magnetic susceptibilities.

The now-conventional Lifshitz theory has been extended to magnetically susceptible media by use of the surface-mode approach [5.78]. This paper determines the free energy of interaction of parallel half-spaces separated by a slab, all three of which may be electrically and magnetically susceptible and electrically conductive. The electrical and magnetic susceptibilities enter in a completely symmetric fashion to generalize earlier work DZYALOSHINSKII et al. [5.58] did for the same geometry for solely electrically susceptible media.

A quite different approach to the van der Waals interaction which employs methods from theoretical elementary particle physics has appeared [5.79,80]. By considering the van der Waals interaction as an elastic two-photon exchange process, they derived a formula for the two-particle potential valid only in the fully retarded case which has been generalized by LUBKIN [5.81] to

$$U(r) = \frac{\hbar c}{4\pi r^7}\left[-23\left(\alpha_E^A\alpha_E^B + \alpha_M^A\alpha_M^B\right) + 7\left(\alpha_E^A\alpha_M^B + \alpha_E^A\alpha_M^B\right) - 60\alpha_X^A\alpha_X^B\right] \quad . \tag{5.40}$$

In this equation, r is the separation of particles A and B, the α's are static polarizabilities, and their subscripts E, M, and X are respectively for electric, magnetic, and crossed polarizabilities (see [5.81] for a discussion of α_X). In their explicit expressions for the interaction, FEINBERG and SUCHER assumed the interparticle separations to be large compared to the particles' sizes, with the consequence that their results are useful only in the near-retarded and fully retarded zones. Since these are generally not the most important zones for aerosol applications with the approximations precluding treatment of the particles upon close approach as resembling interacting slabs (see Sect.5.4.6), considerable additional work is needed. Nonetheless it would be useful to compare results of RICHMOND and NINHAM [5.78] suitably generalized to spheres in the retarded zone with (5.37) and other results of this approach as an indication of both possible asymptotic forms and omitted terms. Another approach also relevant only to the fully retarded interaction considers the perturbation of the zero-point energy of the electromagnetic field. BOYER [5.82,83] has employed this approach to derive the fully retarded interaction potential for an electrically and magnetically polarizable particle and a perfectly conducting wall

$$U(r) = \frac{3\hbar c}{8\pi r^4}(\alpha_E - \alpha_M) \quad . \tag{5.41}$$

Though the domain of applicability for aerosols of (5.41,42) and other formulas for the fully retarded interaction is to be questioned because of both the uncertainties in the force due to long-wavelength contributions and the mean free path for the particle, they do point to potentially significant effects. For example, the repulsive interaction between a magnetically susceptible particle with low electrical susceptibility and a conducting surface could play an important role in the anomalously low removal rates observed by ZAGAINOV et al. [5.21].

5.4.6 Physical Effects and the Application to Real Systems

The discussion of van der Waals forces has thus far been primarily concerned with the interactions of infinite, homogeneous half-spaces at absolute zero on the thermodynamic temperature scale. In applications to aerosols, these results serve as the starting point in deriving expressions of use for describing the interactions of finite bodies either with each other or with realistic media, either or both of which may be layered because of condensational formation processes and exhibit finite temperature effects.

Inclusion of Temperature Dependence

The van der Waals interaction was introduced earlier in this chapter as the difference between the zero-point energies $[(1/2)\hbar(\omega - \Omega)$ or (5.30)] of the interacting bodies at infinite and finite separations. NINHAM et al. [5.84] indicated how these results should be rigorously extended to finite temperatures without resorting to temperature Green's function methods of quantum-field theory as was originally done by DZYALOSHINSKI et al. [5.58]. They generalized the harmonic-oscillator zero-point energy above by considering the Helmholtz free energy of the harmonic oscillator at finite temperature $G = kT \sum_j \ln[2 \sinh(h\omega_1/2kT)]$ were ω_j are the oscillator eigenfrequencies. The ω_j characterize the interacting media and are given by their dispersion relations, $D(\omega) = 0$ [e.g., (5.36)]. By a procedure similar to (5.32,33) and with care in performing the complex integrals involved, expressions for the free energy of interaction necessarily including temperature dependence may be obtained.

The importance of temperature in lipid-water and other systems was discussed by PARSEGIAN and NINHAM [5.85]. They argued that due to the peculiarities of water, the temperature-dependent intermolecular correlations are always important.

For small particles, inclusion of temperature effects related strictly to the particles' oscillator modes is generally unnecessary whenever the thermal wavelength is much greater than the largest dimension of the particle, $\lambda_T \equiv (hc/kT) >> d$. This is due to the inability of such a cavity (i.e., particle) to support all possible modes as does a body which is infinite in one or more dimensions [5.86]. However, the continuum-mode density must be included for those nonfinite media with which the particle interacts. Thus, for example, a 10 nm water particle near a water surface covered with lipid film (to use [5.85] chemical constituents) could respond to the thermal emissions of the lipid-water half-space as its electric susceptibility dictated, but it would have no thermal emissions of its own to contribute to the interaction energy of the system.

Geometry and van der Waals Interaction

The traditional approach to van der Waals force calculations employing Hamaker theory is illustrated in a series of papers [5.87-91] discussing anisomeric and layered-particle interactions. These papers give numerous explicit results including both indication of the conditions for repulsive interaction and the effects of geometry on the interaction forces. Their formulas are useful for qualitative purposes, but the limitations of the method raise questions both as to their general validity and certainly as to their potential for quantitative predictions.

For the understanding and description of aerosol particle interaction processes, the two most important questions are what the two-particle interaction forces are and what the particle-surface interaction forces are. Since a substantial fraction of condensational aerosol particles from all sources nucleate to form spheres prior to subsequent coagulation and deposition, the description of the interaction of spheres is of central interest.

For the case of interacting homogeneous spheres, characterizations of the interaction energy which are valid in various limiting cases have been derived. By computing the vacuum modes alone (i.e., the material-independent component of the interaction) LANGBEIN [5.92] showed a quite varied interaction energy depending mutually upon the relative dimensions of 1) interacting particle radii, 2) wavelength of the interaction frequency under examination, and 3) particle separation distance. By themselves, these results do not describe interactions among material media, but they do illustrate the role geometry plays. Explicit upper and lower limits on the nonretarded interaction energy have been derived from its full series expansion [5.93,94] and will be discussed below in the more general context of multilayered spheres where approximations prove more useful. In all cases the interaction energy does vary from the form first derived by LIFSHITZ for interacting half-spaces at close approach to the conventional nonretarded $1/d^6$ form for the interaction of spheres. A readily calculable approximation to Langbein's results for identical homogeneous spheres of dielectric permeability ε_s interacting across a medium of permeability ε_m was found by KEIFER et al. [5.95]. Their solution is for the nonretarded case of two spheres of radii R whose center-to-center separation is ZR. They defined $\varepsilon(\omega) = \varepsilon_s(\omega)/\varepsilon_m(\omega)$, $z = 2\cosh(\theta)$, and $\xi_n = 4\pi^2 kT/h)n$ for n = 0, 1, 2, ... The interaction free energy is

$$G_s(Z) = -kT \sum_{n=0}^{\infty}{}' g(i\xi_{nj};Z) \tag{5.42}$$

where the ' indicates the n = 0 term is multiplied by 1/2. Here,

$$\begin{aligned} g(i\xi;Z) = \frac{1}{8}\sum_{\nu=1}^{\infty} &\left[\frac{1}{\sinh^2(\nu\theta)} + \frac{1}{\cosh^2(\nu\theta)}\right]\frac{Q^{2\nu}(\xi;Z)}{\nu} \quad , \\ &-\ln\left\{\left[1 + F[Q(\xi;n)]\right]\left[1 + F[-Q(\xi;Z)]\right]\right\} \end{aligned} \tag{5.43}$$

$$Q(\xi;Z) = [\varepsilon(i\xi) - 1]/[\varepsilon(i\xi) + \beta(Z)] \quad , \tag{5.44}$$

$$\beta(Z) = 2[a(Z)/b(Z)]^{\frac{1}{2}} - 1 \quad , \tag{5.45}$$

$$a(Z) = \frac{1}{2}\left(\frac{1}{Z^2-4} + \frac{3}{Z^2}\right) - \frac{2}{Z^2-1} \tag{5.46a}$$

$$b(Z) = \frac{1}{2}\left(\frac{1}{Z^2-4} + \frac{1}{Z^2}\right) + \frac{1}{4}\ln\left(1 - \frac{4}{Z^2}\right) \tag{5.46b}$$

$$F(Q) = \sum_{m=1}^{\infty} \frac{\sinh\theta}{\sinh(m+1)\theta} Q^m \quad . \tag{5.47}$$

The authors find very rapid convergence of (5.42). Model calculations indicated agreement with the exact formula to 2 percent.

Many aerosols such as chain agglomerates and fibers may be viewed, to a first approximation, as cylinders. Interactions between parallel cylinders have been studied by LANGBEIN [5.96] who found interaction laws between those for interacting spheres and interacting half-spaces for nonretarded and presumably, therefore, for retarded cases. For separations respectively small and large with respect to cylinder radii, nonretarded interaction energy laws varying as $d^{-3/2}$ and d^{-5} were computed. MITCHELL and NINHAM [5.97] have calculated the nonretarded interaction between thin cylinders inclined at an angle enabling alignment torques to be computed.

Nonretarded van der Waals forces involving anisotropic ellipsoidal particles have been calculated by IMURA and OKANO [5.98] assuming separations much greater than characteristic particle sizes. They found that depending upon relative values of the particle's dielectric permeabilities along its different axes, preferential alignment occurred. This was of import for both particle-particle and particle-wall interactions.

Interactions among lamellar particles and surfaces are important in many contexts. Most surfaces have a contaminant or oxide layer, while particles can be built up from a primary nucleus which subsequently undergoes a variety of thermodynamic conditions. In this fashion, one or more layers of potentially different composition from the nucleus can be added to form a lamellar particle. Interactions of multilayer systems of any geometry resemble the interactions between half-spaces for very small separations. DZYALOSHINSKII et al. [5.58] gave the expression for the nonretarded, zero temperature interaction force between two half-spaces of frequency-dependent dielectric susceptibilities $\varepsilon_1(\chi)$ and $\varepsilon_2(\chi)$ which were separated by a slab of susceptibility $\varepsilon_3(\chi)$ and thickness ℓ as

$$F = \frac{\hbar}{8\pi^2\ell^3}\int_0^\infty \frac{[\varepsilon_1(i\chi)-\varepsilon_3(i\chi)][\varepsilon_2(i\chi)-\varepsilon_3(i\chi)]}{[\varepsilon_1(i\chi)+\varepsilon_3(i\chi)][\varepsilon_2(i\chi)+\varepsilon_3(i\chi)]}\,d\chi \quad . \tag{5.48}$$

These results have been extended to half-spaces of arbitrary numbers of layers by PARSEGIAN and NINHAM [5.99] and LANGBEIN [5.100]. The interaction forces among any layered systems upon close approach should be essentially proportional to F in (5.48). It is clear that F may be only attractive when the intervening medium is a vacuum ($\varepsilon_3 = 1$). To the contrary, an example of a repulsive interaction was provided by RICHMOND et al. [5.101] who calculated the chemical potentials for five systems each consisting of water "wetted" by a thin film of hydrocarbon with air (or a vacuum) above it. In agreement with experiment, they found negative total chemical potentials for pentane, hexane, and heptane on the water surface indicating film spreading whereas octane's and dodecane's chemical potentials were positive indicating film thickening. Layering was addressed in some detail for spheres in their nonretarded region by LANGBEIN [5.93,94]. He found the dispersion energy to be determined by the outer layer for particle separations less than the layer thickness and by the nucleus for greater separations. When a multilayered particle was considered, LANGBEIN [5.34] correspondingly argued that all layers need be considered for particles a large distance apart, whereas this complicated dependence diminishes in favor of the outer layers as the particles approach.

The interaction between a homogeneous sphere and surface is available from the general results of interacting spheres by allowing the radius of one sphere to go to infinity. KEIFER et al. [5.95] have given another easily calculable approximation to Langbein's infinite-series expression for the interaction energy [5.94]. In their notation, ρ is the radius of the sphere divided by the distance from the wall to the sphere's center. Their θ is defined by $1/\rho = \cosh 2\theta$. The dielectric permeabilities of the sphere, wall, and intervening medium are respectively $\varepsilon_s(\omega)$, $\varepsilon_w(\omega)$, and $\varepsilon_m(\omega)$. In these variables, $\varepsilon_1 = \varepsilon_w/\varepsilon_m$, $\varepsilon_2 = \varepsilon_s/e$, and $Q = [(\varepsilon_1 - 1)/(\varepsilon_1 + 1)] \cdot [(\varepsilon_2 - 1)/(\varepsilon_2 + 1)]$. Their interaction energy between wall and sphere is

$$\Delta E_{ws} \cong - \frac{\hbar}{32\pi} \int\limits_{-\infty}^{+\infty} d\omega \left\{ \sum_{\nu=1}^{\infty} \left[\frac{1}{\sinh^2(\nu\theta)} + \frac{1}{\cosh^2(\nu\theta)} \right] \frac{Q^\nu}{\nu} \right.$$

$$\left. - \ln \sum_{m=0}^{\infty} \frac{\sinh(2\theta)}{\sinh^2(m+1)\theta} Q^m \right\} \tag{5.49}$$

where Q is defined as follows

$$E_1 = \frac{1}{2}\left(\frac{\rho}{1-\rho^2} - \tanh^{-1}\rho\right) \tag{5.50a}$$

$$E_2 = \frac{1}{2}\left(\frac{\rho^3}{1-\rho^2}\right) \tag{5.50b}$$

$$\chi = \frac{1}{2} \sum_{n=1}^{\infty} \frac{n}{n+(\varepsilon_2+1)^{-1}} \rho^{2n+1} \tag{5.51}$$

$$Q = Q_0 \frac{\chi}{E_2} \quad . \tag{5.52}$$

Model calculations done by KEIFER et al. [5.95] indicated agreement with Langbein's series result to within 1 percent.

Recently, KIEFER et al. [5.102] have published their relatively easily calculated approximations of the general homogeneous sphere-sphere and sphere-wall interactions that simplify the results of LANGBEIN [5.34,94].

RICHMOND [5.103] and IMURA and OKANO [5.98] have discussed the nonretarded interaction of a rod (or ellipsoid) and a half-space. The more restricted cases of point particles interacting with a half-space has been given by PARSEGIAN [5.40] and of a point particle with a multilayered half-space, as in physical absorption, by MAHANTY and NINHAM [5.104].

VAN BREE et al. [5.105] estimated the influence of surface irregularities, as occur in realistic cases, on the van der Waals interaction. Their conclusion was that it is of principal importance in nonretarded interactions where it gave rise to 50 and 25 percent effects, respectively, in the plane-plane and plane-sphere examples they examined.

Another class of interfacial imperfections which affects particle interaction energy is inhomogeneity of the dielectric properties of films on substrates. WEISS et al. [5.106] examined this question for covered half-spaces interacting across a slab and found that accommodation of spacial variations in dielectric susceptibility was important when the distance between bodies was comparable to the region over which the susceptibility was a continuously varying function. In a subsequent paper, KIEFER et al. [5.107] examined the effect of dielectric inhomogeneity of a surface layer (whose dielectric properties were fairly similar to those of the substrate) upon the interaction energy of two spheres. Their model calculation indicated a diminution in the interaction energy compared with the bare-sphere case.

5.5 Summary and Conclusion

This survey has attempted to present the subject of aerosol interaction forces in its full physical context. It was, therefore, necessary from the outset to identify and delineate the numerous components of the aerosol which contribute in an essential way to microphysical aerosol interaction forces. These components include gas composition and density and particle structure and composition, that is, gas dynamics and chemical physics. Though the former is the subject of Chap.2 of this volume, the state of current practice in that area as regards kinetic theoretical definition of the importance of the chemical physics in aerosol interactions was critically assessed here to establish the necessity of the subsequent discussion. A model calculation was discussed which attempted to include aspects of both finite Knudsen number transport and realistically complicated interaction potentials. The conclusion was that the *combined* effects of gas and particle properties are essential for the description of aerosol interactions.

Multipolar electrostatic interactions in the aerosol milieu were concluded not to be of general importance while induced electrostatic interactions possibly were. In this latter case, a calculation from a recent paper was discussed and some useful formulas presented.

The most commonly overlooked aerosol interaction force is that between neutral particles having no permanent multipole moments, the van der Waals force. The conclusions of the earlier discussions of aerosol kinetic theory, both from the views of current practice and the model calculation, indicated the importance of the *structure* of the particle interaction potential at separation distances below a gas molecular mean free path for all transition-regime and smaller aerosol particle interactions. Since this is the domain of the van der Waals potential, the largest single section of Chap.5 was devoted to an introductory discussion of developments of the last decade in this field. Observations on the usefulness of aspects of the theory for aerosol considerations were made and some useful, recently derived formulas were presented.

Acknowledgments

The author gratefully acknowledges useful conversations with Dr. Michael Creutz of Brookhaven National Laboratory and the numerous helpful suggestions by Dr. V.A. Parsegian of the National Institute of Health on the manuscript of this survey.

This survey was supported under the auspicies of the United States Department of Energy under Contract No. EY-76-C-02-0016.

References

5.1 N.A. Fuchs: *The Mechanics of Aerosols*, transl. by R.E. Daisley and Marina Fuchs (Pergamon, Oxford 1964)
5.2 G. Zebel: "Coagulation of Aerosols", in *Aerosol Science*, ed. by C.N. Davies (Academic Press, New York 1966)
5.3 G.M. Hidy, J.R. Brock: *The Dynamics of Aerocolloidal Systems* (Pergamon, Oxford 1970)
5.4 S.G. Ash, D.H. Everett, C. Radke: J. Chem. Soc., Faraday Trans. 2 *69*, 1256-1277 (1973)
5.5 Y.S. Sedunov: *Physics of Drop Formation in the Atmosphere*: transl. by D. Lederman (Wiley, New York 1974)
5.6 H. Margenau, N.R. Kestner: *Theory of Intermolecular Forces*, 2nd ed. (Pergamon, Oxford 1971)
5.7 G. Biczo, S. Suhai: Phys. Lett. *51*A, 223-225 (1975)
5.8 J.R. Brock: J. Appl. Phys. *41*, 843-844 (1970)
5.9 W.H. Marlow, J.R. Brock: J. Colloid Interface Sci. *50*, 32-38 (1975)
5.10 J. Phys. *38*, Colloq. C-2 (1977)
5.11 W.H.Marlow (ed.): *Aerosol Microphysics II: Chemical Physics of Microparticles*, Topics in Current Physics (Springer, Berlin, Heidelberg, New York) in preparation
5.12 U. Kreibig: J. Phys. F*4*, 999-1014 (1974)
5.13 J.R. Brock: J. Colloid Interface Sci. *39*, 418-420 (1972)
5.14 W.H. Marlow: "Atmospheric Electricity and Air Pollution, Ph.D. Thesis, University of Texas at Austin (1974)
5.15 W.H. Marlow: J. Colloid Interface Sci. *64*, 543-548 (1978)
5.16 W.H. Marlow: J. Colloid Interface Sci. *64*, 549-554 (1978)
5.17 N.D. Lang, W. Kohn: Phys. Rev. B*7*, 3541-3550 (1973)
5.18 F.P. Buff, N.S. Goel, J.R. Clay: J. Chem. Phys. *63*, 1367-1379 (1975)
5.19 F.P. Buff, N.S. Goel: J. Chem. Phys. *56*, 2405 (1972)
5.20 D.L. Mills: Phys. Rev. B *15*, 763-770 (1977)
5.21 V.A. Zagainov, A.G. Sutugin, I.V. Petryanov-Sokolov, A.A. Lushinikov: Dokl. Akad. Nauk SSSR *221*, 239-241 (1975); J. Aerosol Sci. *7*, 389-395 (1976)
5.22 D. Sinclair: Am. Ind. Hyg. Assoc. J., November, 729-735 (1972)

5.23 W.H. Marlow, R.L. Tanner: Anal. Chem. *48*, 1999-2001 (1976)
5.24 S.K. Friedlander: *Smoke, Dust and Haze* (Wiley, New York 1977)
5.25 D. Langbein: J. Phys. Chem. Sol. *32*, 133-138 (1971)
5.26 J.N. Israelachvili: Proc. R. Soc. London A *331*, 39-55 (1972)
5.27 A.D. McLachlan: Proc. R. Soc. A *271*, 387-401 (1963)
5.28 A.D. McLachlan: Proc. R. Soc. A *274*, 80-90 (1963)
5.29 A.D. McLachlan: Mol. Phys. *6*, 423-427 (1963)
5.30 A.D. McLachlan: Mol. Phys. *7*, 381-388 (1963)
5.31 E.M. Lifshitz: Sov. Phys. JETP *2(1)*, 73-83 (1956)
5.32 H.B. Callen, T.A. Welton: Phys. Rev. *83(1)*, 34-40 (1951)
5.33 R. Kubo: Rep. Prog. Phys. *29*, part 1, 255-284 (1966)
5.34 D. Langbein: *Theory of Van der Waals Attraction*, Springer Tracts in Modern Physics, Vol.12 (Springer, Berlin, Heidelberg, New York 1974)
5.35 J. Mahanty, B.W. Ninham: *Dispersion Forces* (Academic Press, Lonon 1976)
5.36 V.A. Parsegian: "Long Range van der Waals Forces", in *Physical Chemistry: Enriching Topics from Colloid and Surface Science*, ed. by H. van Olphen, K.J. Mysels (Theorex, La Jolla 1975)
5.37 Y. Yasuda: J. Phys. Jpn. *25* (3), 721-728 (1968)
5.38 V.A. Parsegian, B.W. Ninham: J. Colloid Interface Sci. *37*, 332-341 (1971)
5.39 E.R. Smith, D.J. Mitchell, B.W. Ninham: J. Colloid Interface Sci. *45*, 55-68 (1973)
5.40 V.A. Parsegian: Mol. Phys. *27*, 1503-1511 (1974)
5.41 J. Mahanty, B.W. Ninham: J. Phys. A *5*, 1447-1452 (1972)
5.42 J. Mahanty, B.W. Ninham: J. Chem. Soc. Faraday Trans. 2 *71*, 119-137 (1975)
5.43 N.G. Van Kampen, B.R.A. Nijboer, K. Schram: Phys. Lett. *26*A, 307-308 (1968)
5.44 E. Gerlach: Phys. Rev. B *4*, 393-396 (1971)
5.45 P. Richmond, B.W. Ninham: J. Phys. C *4*, 1988-1993 (1971)
5.46 B.R.A. Nijboer, M.J. Renne: Phys. Norvegica *5*, 243-251 (1971)
5.47 K. Schram: Phys. Lett. *43*A, 282-284 (1973)
5.48 D. Langbein: J. Chem. Phys. *58*, 4476-4481 (1973)
5.49 B. Davies: Chem. Phys. Lett. *16*, 388-390 (1972)
5.50 L.D. Landau, E.M. Lifshitz: *Electrodynamics of Continuous Media* (Pergamon Press, Oxford 1960)
5.51 H.R. Philipp, E.A. Taft: Phys. Rev. *113*, 1002-1005 (1959)
5.52 E.A. Taft, H.R. Philipp: Phys. Rev. *138*, A197-A202 (1965)
5.53 H.R. Philipp: Phys. Rev. *16*, 2896-2900 (1977)
5.54 C.G. Granquist, R.A. Burhman, J. Wyns, A.J. Sievers: Phys. Rev. Lett. *37*, 625-629 (1976)
5.55 U. Kreibig: Z. Phys. *234*, 307-318 (1970)
5.56 B.V. Derjaguin, I.I. Abrikosova, E.M. Lifshitz: Q. Rev. (London) *10*, 295-329 (1956)
5.57 B.W. Ninham, V.A. Parsegian: Biophys. J. *10*, 646-663 (1970)
5.58 I.E. Dzyaloshinskii, E.M. Lifshitz, L.P. Pitaevskii: Adv. Phys. *10*, 165-209 (1961)
5.59 L. Genzel, T.P. Martin, U. Kreibig: Z. Phys. B*21*, 339-346 (1975)
5.60 J.N. Israelachvili, D. Tabor: Proc. R. Soc. London *331*, 19-38 (1972)
5.61 E.S. Sabisky, C.H. Anderson: Phys. Rev. A *7*, 790-806 (1973)
5.62 A. Shih: Phys. Rev. A *9*, 1507 (1974)
5.63 A. Shih, V.A. Parsegian: Phys. Rev. A *12*, 835-841 (1975)
5.64 S. Norman, T. Anderson, C.G. Granquist, O. Hunderi: Phys. Rev. B*18*, 674-695 (1978)
5.65 J.N. Israelachvili, D. Tabor: "Van der Waals Forces: Theory and Experiment", in *Progress in Surface and Membrane Science*, Vol.7, ed. by D.A. Cadenhead, J.F. Danielli, M.D. Rosenberg (Academic Press, New York 1973)
5.66 W.A.B. Donners, J.B. Rijnbout, A. Vrij: J. Colloid Interface Sci. *61*, 535-544 (1977)
5.67 R.A. Craig: J. Chem. Phys. *58*, 2988-2993 (1973)
5.68 J. Heinrichs: Phys. Rev. B *11*, 3625-3636 (1975)
5.69 J.E. Inglesfield, E. Wikborg: J. Phys. F *5*, 1475-1489 (1975)
5.70 J. Harris, R.O. Jones: J. Phys. F *4*, 1170-1186 (1974)
5.71 G.D. Mahan: Phys. Rev. B *12*, 5585-5589 (1975)

5.72 E. Zaremba, W. Kohn: Phys. Rev. B *13*, 2270-2285 (1976)
5.73 D. Chan, P. Richmond: J. Phys. C *9*, 153-162 (1976)
5.74 V.A. Parsegian: "Long-Range Physical Forces in the Biological Milieu", in *Annual Review of Biophysics and Bioengineering*, ed. by L.J. Mullins, W.A. Hagins, L. Stryer (Annual Review Inc., Palo Alto, 1973)
5.75 J. Hubbard, L. Onsager: J. Chem. Phys. *67*, 4850-4857 (1977)
5.76 B. Davies, B.W. Ninham: J. Chem. Phys. *56*, 5797-5801 (1972)
5.77 C.J. Barnes, B. Davies: J. Chem. Soc., Faraday Trans. 2 *71*, 1667-1689 (1975)
5.78 P. Richmond, B.W. Ninham: J. Phys. C *4*, 1988-1991 (1971)
5.79 G. Feinberg, J. Sucher: Phys. Rev. A *2*, 2395-2415 (1970)
5.80 C.-K.E. Au, G. Feinberg: Phys. Rev. A *6*, 2433-2451 (1972)
5.81 E. Lubkin: Phys. Rev. A *4*, 416-419 (1971)
5.82 T.H. Boyer: Phys. Rev. *180*, 19-24 (1969)
5.83 T.H. Boyer: Phys. Rev. A *9*, 2078-2084 (1974)
5.84 B.W. Ninham, V.A. Parsegian, G.H. Weiss: J. Stat. Phys. *2*, 323-328 (1970)
5.85 V.A. Parsegian, B.W. Ninham: Biophys. J. *10*, 664-674 (1970)
5.86 H.P. Baltes, E.R. Hilf: *Spectra of Finite Systems* (Bibliographisches Institut-Wissenschaftsverlag, Mannheim, Wien, Zürich 1976)
5.87 M.J. Vold: J. Colloid Interface Sci. *9*, 541-549 (1954)
5.88 M.J. Vold: Proc. Indian Acad. Sci. Sect. A *46*, 152-166 (1957)
5.89 M.J. Vold: J. Colloid Interface Sci. *16*, 1-12 (1961)
5.90 D.W.J. Osmond, B. Vincent, F.A. Waite: J. Colloid Interface Sci. *42*, 262-269 (1973)
5.91 B. Vincent: J. Colloid Interface Sci. *42*, 270-285 (1973)
5.92 D. Langbein: Phys. Rev. B *2*, 3371-3383 (1970)
5.93 D. Langbein: J. Adhes. *1*, 237-245 (1969)
5.94 D. Langbein: J. Phys. Chem. Sol. *32*, 1657-1667 (1971)
5.95 J.E. Kiefer, V.A. Parsegian, G.H. Weiss: J. Colloid Interface Sci. *57*, 580-582 (1976)
5.96 D. Langbein: Phys. Kondens. Materie *15*, 61-86 (1972)
5.97 D.J. Mitchell, B.W. Ninham: J. Chem. Phys. *59*, 1246-1252 (1973)
5.98 H. Imura, K. Okano: J. Chem. Phys. *58*, 2763-2776 (1973)
5.99 V.A. Parsegian, B.W. Ninham: J. Theor. Biol. *38*, 101-109 (1973)
5.100 D. Langbein: J. Adhes. *6*, 1-13 (1974)
5.101 P. Richmond, B.W. Ninham, R.H. Ottewill: J. Colloid Interface Sci. *45*, 69-80 (1973)
5.102 J.E. Kiefer, V.A. Parsegian, G.H. Weiss: J. Colloid Interface Sci. *67*, 140-153 (1978)
5.103 P. Richmond: J. Chem. Soc., Faraday Trans. 2 *70*, 229-239 (1974)
5.104 J. Mahanty, B.W. Ninham: J. Chem. Soc. Faraday Trans. 2 *70*, 637-650 (1974)
5.105 J.L. Van Bree, J.A. Poulis, B.J. Verhaar, K. Schram: Physica *78*, 187-190 (1974)
5.106 G.H. Weiss, J.E. Kiefer, V.A. Parsegian: J. Colloid Interface Sci. *45*, 615-625 (1973)
5.107 J.E. Kiefer, V.A. Parsegian, G.H. Weiss: J. Colloid Interface Sci. *51*, 543-546 (1975)

References Added in Proof

W. Arnold, S. Hunklinger, K. Dransfeld: Phys. Rev. *19*, 6049-6056 (1979)
W.H. Marlow: "Lifshitz-van der Waals forces in aerosol collisions, I: Introduction; water droplets" (submitted for publication)
W.H. Marlow: "Derivation of aerosol collision rates for singular attractive contact potentials" (submitted for publication)

Subject Index

H. Haken

Synergetics

An Introduction

Nonequilibrium Phase Transitions and Self-Organization in Physics, Chemistry and Biology
Springer Series in Synergetics
2nd enlarged edition. 1978. 152 figures, 4 tables. XII, 355 pages
ISBN 3-540-08866-0

Contents:
Goal. – Probability. – Information. – Chance. – Necessity. – Chance and Necessity. – Self-Organization. – Physical Systems. – Chemical and Biochemical Systems. – Applications to Biology. – Sociology: A Stochastic Model for the Formation of Public Opinion. – Chaos. – Some Historical Remarks and Outlook.

Laser Monitoring of the Atmosphere

Editor: E. D. Hinkley
1976. 84 figures. XV, 380 pages
(Topics in Applied Physics, Volume 14)
ISBN 3-540-07743-X

Contents:
E. D. Hinkley: Introduction. – *S. H. Melfi:* Remote Sensing for Air Quality Management. – *V. E. Zuev:* Laser-Light Transmission through the Atmosphere. – *R. T. H. Collis, P. B. Russell:* Lidar Measurement of Particles and Gases by Elastic Backscattering and Differential Absorption. – *H. Inaba:* Detection of Atoms and Molecules by Raman Scattering and Resonance Fluorescence. – *E. D. Hinkley, R. T. Ku, P. L. Kelley:* Techniques for Detection of Molecular Pollutants by Absorption of Laser Radiation. – *R. T. Menzies:* Laser Heterodyne Detection Techniques.

Monte Carlo Methods

in Statistical Physics

Editor: K. Binder
1979. 91 figures, 10 tables. XV, 376 pages
(Topics in Current Physics, Volume 7)
ISBN 3-540-09018-5

Contents:
K. Binder: Introduction: Theory and "Technical" Aspects of Monte Carlo Simulations. – *D. Levesque, J. J. Weis, J. P. Hansen:* Simulation of Classical Fluids. – *D. P. Landau:* Phase Diagrams of Mixtures and Magnetic Systems. – *D. M. Ceperley, M. H. Kalos:* Quantum Many-Body Problems. – *H. Müller-Krumbhaar:* Simulation of Small Systems. – *K. Binder, M. H. Kalos:* Monte Carlo Studies of Relaxation Phenomena: Kinetics of Phase Changes and Critical Slowing Down. – *H. Müller-Krumbhaar:* Monte Carlo Simulation of Crystal Growth. – *K. Binder, D. Stauffer:* Monte Carlo Studies of Systems with Disorder. – *D. P. Landau:* Applications in Surface Physics.

Picture Processing and Digital Filtering

Editor: T. S. Huang
2nd corrected and updated edition. 1979
113 figures, 7 tables. XIII, 297 pages
(Topics in Applied Physics, Volume 6)
ISBN 3-540-09339-7

Contents:
T. S. Huang: Introduction. – *H. C. Andrews:* Two-Dimensional Transforms. – *J. G. Fiasconaro:* Two-Dimensional Nonrecursive Filters. – *R. R. Read, J. L. Shanks, S. Treitel:* Two-Dimensional Recursive Filtering. – *B. R. Frieden:* Image Enhancement and Restoration. – *F. C. Billingsley:* Noise Considerations in Digital Image Processing Hardware. – *T. S. Huang:* Recent Advances in Picture Processing and Digital Filtering. – Subject Index.

Springer-Verlag Berlin Heidelberg New York